Directing Corporate Video

Other Focal Press books by Ray DiZazzo

Corporate Scriptwriting: A Professional's Guide

Corporate Television: A Producer's Handbook

Directing Corporate Video

Ray DiZazzo

Photography by Chris Gerlach

Focal Press
Boston London

Focal Press is an imprint of Butterworth–Heinemann.

∞ Recognizing the importance of preserving what has been written, it is the policy of Butterworth–Heinemann to have the books it publishes printed on acid-free paper, and we exert our best efforts to that end.

Growth and survival tips reproduced with consent.

Library of Congress Cataloging-in-Publication Data

DiZazzo, Raymond.
Directing corporate video / Ray DiZazzo ; photography by Chris Gerlach.
p. cm.
Includes bibliographical references and index.
ISBN 0-240-80164-4
1. Video recordings—Production and direction. 2. Industrial television. I. Title.
PN1992.94.D585 1993
791.45′0233—dc20 92-34624
CIP

British Library Cataloguing-in-Publication Data

A catalogue record for this book is available from the British Library.

Butterworth–Heinemann
80 Montvale Avenue
Stoneham, MA 02180

10 9 8 7 6 5 4 3 2 1

Printed in the United States of America

I would like to dedicate this book to the many crew members, editors, and actors and actresses who have generously given their time, skill, and energy to help make my work successful. I am also grateful to Jenny Norman for her help with the manuscript and to Chris Gerlach for his photographic work and general support.

Contents

Preface

I entered corporate television in 1976 as a scriptwriter. At that point in my career, my knowledge of the film and television production process was virtually nonexistent.

Five years later, however, I was lucky enough to land a job writing full-time in a large corporate television facility. This meant that I was suddenly able to see firsthand all the efforts and activities that went into the production of my scripts.

With amazement, I saw sets designed and constructed; lighting plotted and put into place; actors auditioned and cast; locations scouted, reserved, and paid for; props and equipment ordered, delivered, and set up; and clients and producers involved in intense meetings and screenings. I carefully watched every person and activity in the production process, and though I found every one fascinating, I was most fascinated by the directors.

It took very little time to see why so many people in the production end of the film and television businesses wanted to direct. The raw materials used by the director, the amount of creative influence he exerted, and the enormous pressures he sometimes had to endure were staggering.

I directed my first corporate program in 1981. The project was a simple two-camera studio shoot consisting of an on-camera host, a series of graphic cards, character-generated titles, and several tabletop product shots.

During the shoot, I sweated, stuttered, stammered, cued the wrong cameras, called the wrong shots, and asked myself time and again what had ever possessed me to put myself through such mental torture. When the shoot was over, however, and the pressure was off, I realized I had made it through a creative experience like none other in my life.

Since those early years, I have been lucky enough to direct many corporate shows. On every one, the thrill of watching a live performance I had directed play out in front of the camera remained intense. Just as exhilarating were the subsequent edit sessions in which series of shots I had, at times, recorded on different days, in different locations, and from different angles were cut together as seamlessly as if they had happened in the same instant.

I remain a writer in my soul, but directing film and television has gained a very special place in my heart.

My intent with this book is to pass on to you what I've learned in the director's chair. You should know at the outset that this knowledge by itself will not make you a director. What it will do, though, is establish a *foundation*—a sturdy framework on which to assemble a series of real-life events that *will* make you a director.

Those events are things like putting in a great deal of time, sweat, and study on this subject; maintaining a constant awareness and sensitivity to people, events, and subtle changes that pass in and out of your life; mustering the drive and determination to keep pursuing your dream even in the face of constant rejections; and finding (perhaps inventing) ways to eventually direct your own first productions.

I have arranged *Directing Corporate Video* into four parts.

Part I deals with defining the directing profession and exploring the differences between directing in the corporate world and the entertainment industry.

Part II explores the basic skills a director places on that knowledge foundation. These are the aesthetic skills he uses constantly. They include the critical aspects of script analysis and a detailed understanding of the human, audiovisual, technical, and general aesthetics involved in any film or videotape production. A complete overview of the production process is also contained in Part II.

Part III establishes a typical production scenario intended to illustrate how the knowledge and skills discussed in Parts I and II are actually applied on the job. The scenario introduces a fictional free-lance corporate director and the type of script placed in the hands of professionals like him every day in thousands of cities across the country.

Part IV wraps things up with some final words of advice meant to actually get you started in your own directing career.

Throughout this book, you will also find a constant sprinkling of what I call growth and survival tips from professionals in the field. These and the occasional case studies should provide you with still another real-world perspective on the director's work.

Last but not least, the appendices contain actual examples of the various types of organizational, business, and creative documents used by the typical corporate director every day.

My hope is that the sum total of this information will act as a kind of door for you—an entrance into the challenging and exciting world of the corporate video director.

Directing Corporate Video

PART I

The Business

The Directing Profession

1

"Action . . . Cut?"

What exactly does a director do?

At first that sounds like a simple question. The average person would probably answer: "A director sits on the set in a chair that has his name on it. When everything is ready, he says 'Action' and the camera rolls. He watches as the actors do the scene. When the scene is over, he says 'Cut' and the camera stops. He gets ups, walks onto the set, and talks to the actors about what he wants them to do differently. Then he sits down and calls 'Action' again."

True. A director decides when the camera rolls and stops, and he works with the actors to shape their performances. The problem with this brief description is this: What happens in a few critical moments on a set or location reflects only a *fraction* of the director's responsibilities. The true scope of his work is vastly broader and more complex.

A Definition

Before we go further, we must closely examine that phrase "vastly broader and more complex." To do so, let's start with a definition of what a director is:

> The person responsible to the producer for carefully guiding and shaping a complex series of human and technical elements during the preparation and recording of film, videotape, audiotape, or digital segments that must later edit together into an effective audiovisual entity.

It's worth our time to examine each part of this statement in more depth. "The person responsible to the producer . . ." means exactly that. In a large sense, the director's chair is where the buck stops with regard to the preproduction and production phases of a program's development.

If his program turns up a series of poor sound recordings, the director cannot blame it on the sound recordist or boom operator. If the video or film images do not meet the technical standards required by the producer, the director cannot blame it on the camera operator or director of photography. If the performances of the actors are weak, the director cannot blame the casting agency. If the shoot goes over budget, he cannot blame the assistant director or production assistant who did the scheduling. In

The director must have a vision—a vision she should be able to bring to life using the tools and techniques of the medium.

—Patti Ryan
Producer/Director

each of these cases, although a subordinate to the director may have goofed, the director must bear the full responsibility for that person's actions.

"Carefully guiding and shaping . . ." are the director's primary jobs. He is continually guiding and shaping all of the elements at his disposal toward an end result that reflects his interpretation of the script he has been charged with recording. "A complex series of human and technical elements . . ." are what he guides and shapes. Let's touch on both.

The *human* elements are the performances of actors, the comfort level of the client, the satisfaction of the producer, and the chemistry between the director and his crew. The *technical* elements are focal lengths, the choice of lenses or filters, the composition of a shot, lighting decisions, and the choice of when, where, and *if* the camera moves. Technical elements also include microphones, mixing consoles, edit controllers, monitors, and many other pieces of equipment used in the production and postproduction processes.

"During the preparation and recording of film, videotape, audiotape, or digital segments that must later edit together into an effective audiovisual entity" takes into account the medium in which the director is working and the edit sessions that follow production. Edit sessions are the "proof of the pudding," in a sense, because it is in the dimly lit edit suite that the director ultimately finds out if everything he shaped and guided according to his own aesthetic vision now blends seamlessly into a successful overall piece of audiovisual work.

The Director as an Artist

The director is also an artist in a very real and exciting sense. The excitement comes from both the process and the raw materials he uses in creating his work.

The painter creates with a canvas and brushes. Her work is made of colors, textures, lines, strokes, and shapes. The musician creates with an instrument. The manifestation of his work is sound. His creations are notes, melodies, rhythms, volumes, paces, and intensities. The writer creates with the printed word. Her creations are exciting images that play out in the minds of her readers.

Just as paintings are made up of individual brush strokes, programs are created out of individual shots, scenes, and acts.

—Ralph Phillips
Writer/Director

The director, however, creates with all the elements of our world. His materials are people, houses, cars, props, wardrobe, locations, emotions, and character traits. His creations are precisely choreographed illusions of reality—pieces of pre-visualized life that unfold not on a canvas, through a speaker, or on a sheet of paper but as *live events* before the eyes of a viewer.

It is that word *live* and the numerous subtleties inherent in it that add such tremendous scope, complexity, and excitement to the director's work.

To change his work, he does not replay a note, remake a brush stroke, or rewrite a passage. The director must change the way people think about and perceive their world. If he feels that a performance is too static, he must work with the actor to find true motivation for movement. This means exploring what drives the scene, how the actors feel about those things, and how those feelings should be manifested on the screen—with a glance? a movement? a word? or a combination of these?

If he feels that a piece of dialogue breaks the sense of reality, he must examine what words are wrong and why. Are they unmotivated? insincere? erroneous? absurd? Then he must change how his scene plays out in order to create and support the right dialogue.

If the framing or composition of a shot is disturbing or uncomfortable, he must ask himself what subtlety is out of balance—the positions of the actors? the movement of the camera? the focal length? the lighting? the background or foreground? He must then determine what pieces of the scene need to be changed to make the shot aesthetically proper for the tone of the work.

Not for the Fainthearted

When these or any of numerous other elements are not all working together to support his illusion, the director is placed under intense pressure. It is his job to make everything work in perfect sync and with complete believability in front of the camera. Thousands of dollars and a tremendous amount of effort are at stake. Crew, cast, and equipment have been paid for. Props have been ordered, created, painted, and positioned. Wardrobe has been decided on, locations have been scouted, reserved, and rented. Sets have been designed and built.

Each of these elements has been acquired largely based on the director's wishes. The director is well aware that clients and producers—the persons paying for the creation of his vision—are waiting in the wings with bated breath for the result.

For the director, it has to work.

The Reward

And the reward comes, when it does work.

When everything is happening exactly as he had envisioned, the director is fulfilled. When the composition has just the proper depth, color, and balance and the actor's lines, movements, and inflections are delivered with complete sincerity and the camera dollies in to just the proper focal length and angle at just the right moment, the sense of satisfaction is incredible. It can only be matched later in the edit suite when every scene recorded, containing those same perfect elements, is assembled into a larger whole—the program itself.

It is at this point that the creation is complete, and, if indeed everything was directed properly, the illusion is perfect. The viewer sits and watches, and it never occurs to her that a camera was present or that the people moving about on the screen are actors or that virtually every element in the frame was painstakingly positioned according largely to one person's wishes. As the scene plays out all the subtleties of the director's work, all his "shaping and guiding," become invisible. They simply dissolve into the fabric of reality created on the screen.

Directors are driven people, and it is these creative elements that drive them—the intense rewards and the agonizing risks of creating art as a live illusion.

If this is your chosen path, you should know that it is often not a calm or emotionally level career. On the contrary, directing is an extremely intense occupation in which radical highs and lows are common.

That's because despite all the director's vision, sensitivity, and efforts, it doesn't always work. Even the best directors sometimes call it wrong. When this happens, all the time, effort, and money do not pay off. During such times, the director must pick himself up, remuster his belief in himself as an artist, and pursue the next project—with the absolute confidence that next time it *will* work.

In the following chapters, we will explore how all of the elements we've been discussing are combined with still another set of skills—the business skills required for an artist like the director to work successfully in the corporate world. Though the two skills—creative and business—may at first seem like an unlikely pair, you will find that they actually fit together quite well in many respects.

To see how and to further explore the director's world, we will start by looking at how he fits into two very different worlds—the Hollywood back lot and the corporate boardroom.

The Director's Market

2

There are three major job markets for the director: *entertainment, commercials,* and *corporate.* Each of these markets breaks down further into more specialized areas. The entertainment market, for instance, is made up primarily of feature films and television. These two areas break down still further into high- and low-budget films and television productions. This same monetary stratification holds true in the commercials market. There are high- and low-budget commercials and directors who specialize in each. The corporate or industrial market, as it's sometimes called, also varies in terms of money. There are corporate programs made for next to nothing and others that are quite elaborate in scope and production value. Like entertainment films and television, the corporate market is also occupied by both staff and free-lance directors.

Though an entire book could easily be written on the director's job markets alone, we will take just enough time to touch briefly on the benefits and drawbacks of these three primary areas.

The Entertainment Market

It is Hollywood's major studios, such as Paramount, Universal, Disney, and Twentieth Century Fox, that produce thousands of films and television programs every year solely for the sake of entertainment. These include human dramas, action adventures, science fiction thrillers, comedies, horror films, suspense thrillers, situation comedies, soap operas, and others. As you might guess, becoming a director in the high-budget feature film or prime-time television market is an enormous career accomplishment.

High-Budget Feature Films

Although many people in the entertainment industry prefer working in television, high-budget feature films are generally considered the pinnacle of a director's career. They are by far the most expensive and elaborately produced media events created. They are also the most impressively presented, and carry the greatest earning potential.

High-budget features are often filmed in exotic locations and include the most innovative stunt and special-effects work in the world. Some high-budget feature directors are allowed go to virtually any lengths to get just that perfect shot. There is also a flip side to this coin, however. Because of their huge monetary requirements, their enormous technical and logistic characteristics, and the fickle nature of the viewing public, high-budget features also have the greatest potential for loss.

To an aspiring director, all this means several things. First and foremost, it means that getting a directing job in feature films is virtually impossible without years of hard work, a good deal of luck, a high-powered agent, and great connections.

The majority of feature film directors are veterans of the industry who have worked at their craft for many years. Along the way, not only have they learned the production process and the directing craft but they have also gotten to know the right people and organizations to help move their careers forward. The few individuals who are able to land major feature film projects at a young age have usually done so with highly impressive first efforts, such as award-winning student films, short features, or independent productions. When these types of initial efforts garner awards at industry film festivals, they sometimes catch the eye of a feature film producer on the lookout for new talent.

Rewards and drawbacks Once a director has established herself in the high-budget feature film market, both the rewards and the drawbacks are substantial. Hollywood's top feature film directors are paid enormous salaries, and a few are treated to every luxury the studios can offer. They also carry the ultimate clout in the entertainment industry. This is because directors are considered the primary creative force behind a feature film. In fact, they are most often thought of as the authors of the films they direct.

One way this is evidenced is in the opening credits of many feature films seen these days. Next time you go to the theater, look for the title "a Mary Smith film," with Mary Smith being, of course, the director's name. Although hundreds of people have had various levels of involvement in a film—some quite significant, such as the writer, editor, and producer—it is often billed as the director's creation.

Even with all these perks, the work of a top Hollywood director also contains a number of significant drawbacks. One is the intense pressures a director must bear when making creative decisions that affect $50-million budgets. There are also the constant travel and the all-consuming work schedules demanded by projects of such magnitude. Hollywood's avarice for market share and cold cash also breeds a personal struggle that feature film directors often refer to: the continual battles to keep their work pure and insulated from commercial revision or tampering.

Television

Commercial television is the second major entertainment market, and it too bears fruitful rewards in terms of status and monetary gain for its best directors. Salaries are often quite lucrative, and the top directors are treated with great care.

In television, however, the challenges a director must face are often considerably less than those of the high-budget feature film director. Budgets are usually substantially lower, and instead of dealing with the logistics of shooting on locations around the world, many television shows are shot in air-conditioned studios.

In addition, on television programs—such as situation comedies (sitcoms) and soap operas—characters have been previously established and fully developed. They are simply placed in a new series of plot predicaments each week.

This, as you might guess, makes the director's job much easier. Instead of having to study and develop a totally new cast of personalities, she rehearses with the existing seasoned cast and focuses on how their performances should be handled in the current plot situation. Following initial read-throughs, the director and the actors walk through or block the scenes on the set. Finally, they record the script in a single day.

Instead of the logistics of developing camera angles shot, at times, on massive and complex scales, the sit-com television director has perhaps four cameras on the studio floor. In programs shot on film, such as "Cheers" and "Designing Women" all cameras roll simultaneously as the actors perform. Later, the different angles are edited together to make up the show.

In live programs, or those shot on videotape, such as soap operas, news shows, and sports events, the director switches the program *live* or *live to tape*. This means "taking" shots from different cameras on the fly as the show is performed.

The Commercials Market

Commercials are as much a part of American television as the evening news. In an effort to catch and hold a viewer's attention for a short, critical period of time, many commercials go to extreme lengths in terms of production value and special effects.

The top directors who create those commercials routinely make thousands of dollars per day. This is because high-budget commercials, like high-budget feature films, often reflect huge investments. Budgets for 30-second commercials can easily range into the hundreds of thousands of dollars, and those made for seven figures are not uncommon. The airtime purchased to broadcast them can be equally as expensive.

Like the high-budget feature film director, a top commercial director might shoot in a foreign country to achieve just the right shot or perhaps spend days lighting and shooting a single shot of a car in a studio. Or she might direct top celebrity actors to gain a specific type of credibility or product identification.

Because the money can be very good and because challenging dramatic scripts do not come along every day, feature film directors often direct high-budget commercials in between feature film projects. As you might guess, this makes competition for such jobs intense.

Commercial directors land work by contacting advertising agencies contracted by the makers of products. Typically, an agency will ask to see a demo reel and résumé before even considering a new director. As you might guess, most agencies get far more submissions than they can hope to use. This means that years of experience are usually required, along with good connections and perhaps a personal agent to communicate those attributes to the right decision makers in the agency.

Low-Budget Productions

Low-budget commercials are much smaller in scope, as are low-budget features. Cable companies often sell low-budget commercial productions to local merchants for a fraction of the cost of their high-budget counterparts. This is sometimes part of a package deal in which the merchant receives both the commercial production and the airtime on cable for a single, cut-rate price.

While such commercials are rarely noted for their production value or quality look, they do make the opportunity to break into directing much more accessible. The same holds true for low-budget feature films: They don't win awards, but they do offer experience and they are more accessible to the new director.

Payment for low-budget commercials and features is usually in keeping with their very small budgets, often in the range of \$100 or \$200 per day.

The Corporate Market

The corporate staff director works with the same types of production processes that the film, television, and commercial directors employ. The differences in corporate television center primarily on *money* and *content.*

Corporate Budgets

Corporate video budgets are typically much smaller than those of commercial television or feature films. While the average feature film budget might be somewhere in the neighborhood of \$20 million, the average corporate or industrial program budget is around \$20,000.

This means that although the *process* is the same, the *scope* of corporate productions is reduced considerably. For instance, a scene that a feature film crew might shoot in 3 days with a crew of 50 will be shot by the corporate crew in 3 hours with a crew of 5. While a feature film director might line up 3 totally different and distant locations for a particular sequence, the corporate director will line up 1 location and find a way to make it look like 3.

In addition, the corporate scriptwriter knows she must script much less elaborate scenes than would be the case in a feature film screenplay. Part of the producer's job is to ensure that this is the case.

Major budget reductions are also accomplished in corporate television by having crew members perform multiple tasks. A director will sometimes operate her own camera or double as a technical director at the switcher. A grip might also serve as a production assistant, a gaffer, and perhaps a makeup assistant. A director of photography may also act as a gaffer or set builder and perhaps a floor manager as well.

> *Remember, all programs for corporate TV are intended to make people* think.
>
> —Donna Chandler
> Director

Content

A second major difference between corporate and entertainment industry directing is the content of the material shot. In Hollywood, films are most often created simply to entertain. In corporate television, programs must also convey information. This means that while they may also be dramatic, funny, or suspenseful, they must be written and directed somewhat differently from the way theatrical productions are.

A feature film or commercial television director, for instance, may be concerned with conveying one emotional message or idea, perhaps the undying love between a man and a woman. She would probably do this by shaping a series of scenes, some dramatic and perhaps some humorous or suspenseful, but all focused on the concept of love.

A corporate director may be required to convey emotional content, but she is usually also charged with communicating information about her company. This means that while a series of purely dramatic performances might fit the bill in a feature film, a series of informational or instructional performances might work better in a corporate program.

The next logical question, then, might be this: Does the shift in emphasis from dramatic to informational content make the corporate director's job less challenging or rewarding? When I first entered corporate television, I thought surely this would be the case. How, I wondered, could a director get excited about shooting a scene that must communicate the four steps involved when a telephone installer encounters a difficult customer?

The moment I was put behind the camera, however, and given the responsibility of making such a scene work credibly and effectively, my whole outlook changed. I found the experience intensely rewarding and challenging. It quickly became clear to me that although my actors might not be communicating some grand emotion, they were still performing, and shaping their performances was a highly creative task.

There was still blocking to develop: Just how would they move in front of the camera? Also tone: Should they deliver their lines with a positive sense? with a negative sense? with ambivalence? with confidence? Then there was lighting: The scene may not have been a dark, dramatically lit love scene, but if it was lit with skill, I found it could help create and support the tone of the message being communicated. It could also turn a flat, uninteresting collection of objects on the set or location into a visually pleasing screen composition.

In short, my respect for directing corporate programs grew immensely with the first few scenes I directed. It continued to grow as the challenges of the scripts I worked with expanded in scope and complexity.

Thus, although high-budget feature films may be the pinnacle of a directing career, corporate directing is certainly a highly rewarding and challenging stepping-stone for the journey up the creative ladder.

The Free-lancer

Many corporate directors are middle management staff employees of large companies. As such, they are provided with a secure paycheck (usually around $40,000 to $50,000 per year) and a solid benefits package.

Another form of corporate director, though, is the free-lancer. The free-lancer is essentially very much like the staff corporate director in terms of job description. A free-lancer differs, however, in two ways: She has more *freedom* and less *security*.

Freedom A free-lance director is hired on a per-job basis. She is brought in when a script is ready to shoot and is paid a daily or flat fee to prepare or *prep* and record it.

When her job is completed, she is on her own again. This means that the independent director can dabble on both sides of the creative fence, so to speak. She can often develop work outside the corporate umbrella, such as directing commercials, cable programs, and sometimes newsmagazine segments, documentaries, or informational shows. Some free-lance corporate directors also do documentary work and low-budget feature films in hope of eventually crossing over into high-budget projects.

Security The negative aspect of the free-lancer's world is the lack of security and benefits.

Directing jobs can be numerous some months and nonexistent others. Medical and dental bills can be quite high these days. An individual retirement account can be extremely difficult to set up and maintain when the rent comes due and the bank account is nearly dry. The free-lance director deals with these drawbacks on a continual basis.

Some free-lance directors also do other types of production work to supplement their income. Many, for instance, also do camera and lighting work, and quite a few edit as well. Both these skills are natural sidelines for a director, since they also form an inherent part of her primary work.

Your Choice

Whichever directing road you choose to pursue, this book can help you get there. Besides reading, studying, and eventually working at your craft, however, the additional qualities you'll need are desire and tenacity. These two character traits seem somehow to naturally produce the third and final critical element in a director's development—the commitment to plain old hard work.

The Corporate Director

3

❑ ❑ ❑ ❑ ❑

Every successful corporate director is knowledgeable about the general *business* environment in which he works, how *media* is developed and used in that environment, and who the key *players* in the program development process are.

To assume that this general business knowledge is merely background or peripheral in terms of the actual process of directing would be a mistake. Knowledge of business organizations and the people and media within their structures is brought into play directly and continually on every corporate directing project. Once learned, it forms one cornerstone of a foundation that is as much an integral part of the overall directing process as setting up a shot with the director of photography or blocking a scene with a cast of actors. An inability to understand or acknowledge this cornerstone has forced more than one corporate director into another line of work.

The Corporate Environment

"Corporate culture" is a phrase that describes the general philosophies of a corporation or large company. Some companies have very conservative corporate cultures. These organizations typically stress traditional values in their corporate policies and procedures. Often they prefer that employees dress and groom themselves conservatively, stay relatively safe or middle-of-the-road in their thinking, and show a willingness to conduct business in highly structured working environments.

Other organizations play by a more relaxed set of rules. In these companies, a more contemporary standard of dress and grooming is often acceptable, as are an increased tendency toward risk management and a less structured set of business operating policies. The key factor for the free-lance corporate director is to recognize that these cultural differences exist in the companies he works for and to be willing to adjust his style as appropriate.

Making this adjustment in business style is critical to the free-lance director because corporate culture is an underlying element in many business decisions. As an example, suppose a company with a very conservative culture wants a video program made and the company's in-house producer brings in a free-lance director. How might a roomful of dark-suited, white-shirted, button-down-collar executives feel if that director stepped into the boardroom with a 6-inch ponytail and wardrobe selections from one of Hollywood's funkiest shops?

The answer is simple. They would feel uncomfortable. They would have a very difficult time relating to such a director simply because he was not like one of them. He would be considered an outsider with nontraditional standards, and conservative culture suggests that the nontraditional is always suspect. Granted, this discomfort might be overcome if the director's credentials and people skills were good enough. Then again, it might also lead to a quick rethinking of the project before he ever got the chance to display those attributes.

Thus, the lesson here is simple. If you choose to work in corporate television, one of your directorial prerequisites is to gain a healthy knowledge and respect for the workings of the business world and adhere to the established cultural guidelines and traditions of the companies you work for.

Corporate Use of Video

Video programs are used in business to *inform, motivate,* and *train* employees. Although elements like entertainment and drama often become a means to that end, this is not always the case.

Some corporate programs are very straightforward, inexpensive, and simple. One example might be a voice-over narrated program with simple visual shots of the subject being covered—perhaps the way a gas meter is installed or the proper method for stocking shelves in a warehouse.

Other programs may be quite elaborate, with large casts, dramatic plots, character development, and complex human issues explored. Examples of such programs are those on topics like substance abuse, stress reduction, and career burnout.

Most corporate programs fall somewhere in between these two extremes. They often consist of a host on camera, short dramatic or humorous vignettes, employee interviews and on-the-job documentation shots, and perhaps an opening or closing message by an executive—sometimes, the company president.

Cassette Distribution

Whatever the type of program, it is most often shot on videotape and then edited, duplicated, and distributed to company VCR locations on videocassettes. Employees meet periodically, often prior to their work shifts, to view such programs. Many times they also have short, related discussions with local supervision or receive associated handout material. Internal company news programs are utilized in this way, as are safety programs and some instructional programs.

Corporate Broadcast Networks

Another use of video programming is live broadcasts from a studio or conference room uplink location to a network of downlink locations spread throughout the company's operating area or perhaps across the nation.

Major companies like Sears, JC Penney, GTE, and Ford Motor Company have used dedicated private broadcast networks as a regular means of employee training and communication. Employees gather in meeting rooms at the scheduled broadcast times and watch a program as it is happening live, just as they might watch the 5:00 P.M. news.

Corporate broadcasts often also have the unique element of interactivity. Special audio lines and dedicated telephone numbers are established between audience members and executives or panels of experts at the uplink location. Employees are thus able to call in, ask questions about the subject being discussed, and receive on-the-spot answers from those most in the know.

In this way, new products may be introduced to an entire sales force, demonstrated, and even discussed on a national scope—all in one showing. In still another situation, a CEO may go on a live broadcast to get out an important message. Using this technology, he is able to reach virtually every employee in the company with a single presentation.

Interactive Training Programs

Still another use of corporate video is in the development of interactive training programs.

In this medium, programs are scripted according to carefully designed, branching sequences. These sequences are typically shot on videotape; transferred to a second medium, such as laser disk; and used at a specially equipped computer workstation. An employee might sign on to the system and be shown information, perhaps on how to drive a skip loader. He would then be given an on-screen test. He would interact with the program by means of the computer keyboard and answer the questions. He could be routed forward in the program only if he passed the test. If he failed, he would be "branched" back to review the same or similar material and would retake the test—until he got it right.

The Key Players

In each of these cases—videocassette production, live broadcast, or interactive laser disk programs—there are two key players a corporate director must know and work with very closely: *producers* and *clients.*

The Corporate Producer

A client comes to a corporate producer with a communication need. It might be a safety representative requesting a program on proper lifting techniques, or a public affairs representative who requires a public service announcement, or perhaps a training curriculum specialist considering a video program to support a part of a course she is developing.

The producer's role The producer usually oversees video productions like these from inception through distribution. This involves a number of important tasks. A producer will hire a writer to develop a script or write the script herself. She will also develop a budget and a general time line and production schedule. She will hire a director or sometimes direct

Directing corporate television can be more difficult than directing commercial television. It takes a highly talented, creative individual who can work around in-house talent, talking-head scripts, minuscule budgets, and last-minute changes—while still delivering work that is professional and thought-provoking.

—Donna Chandler
Director

the program herself, and she will oversee all business aspects of the project, such as managing the budget, hiring a crew, renting equipment, securing locations, and paying for permits and insurance. Finally, she will hire the editors required on the project and oversee the editing process.

In this role, the corporate producer becomes a unique breed. She must possess and apply both the *creative* skills of a free-lance video producer (often writer and director as well) and the *business* skills of a dedicated company management employee.

Creative versus business decisions In order to continually walk this creative/business "tightrope," the corporate producer must sometimes make judgments the director may not agree with. For instance, a director may feel that opening his program with a short dramatic vignette would be more interesting than the talking-head host opening the writer has scripted, and he might be aesthetically correct in this assumption. The producer, however, may have a number of larger issues to weigh when considering the director's request to change the script. For instance, what will the client think? She and perhaps several other key managers have already approved the script as it stands. Is it now a bit risky to start changing elements just as the project moves into preproduction? Then there's the budget. A dramatic vignette may be the preferred choice on everyone's part, but is the extra money available for actors, additional props, locations, and the like? The issues of deadlines and scheduling also come into play. Will the added shooting and editing delay the release? Finally, there is also the producer's personal opinion. Does she agree that a vignette is the right way to open the show?

After considering all these factors (some creative and some business), the producer may agree to make the change or she may thank the director for his input but decide to move ahead as planned. The prudent director understands this dual role the producer must play. That's because as part of his directing skills he has developed a keen sense of empathy and respect for the *producer's* position, and he knows that it is often far better to concede than to put either the program or his future at this company at risk.

Program work loads Corporate producers typically manage multiple projects at one time. A producer might have two programs in different phases of script development, one in preproduction, one in production, and two in editing.

This means that many producers tend to make decisions that will result in the fewest complications or headaches. Some of those decisions relate to the people the producer hires; others relate to the types of creative elements she allows or disallows in her programs.

For example, the producer will not continue to hire a writer who does inadequate research, because those scripts will often result in client dissatisfaction, last-minute rewrites, or even reshoots. Likewise, she may not bring back a director who continually leaves loose ends, such as poor music selections, incomplete title lists, or hard-to-edit coverage.

The wise director learns to understand this "no loose ends" aspect of the producer's perspective too and to continually manage his own projects so that they move as quickly and painlessly as possible through a producer's schedule.

Clients

Without clients, corporate directors do not work. With clients, however, directors sometimes wish they had never been hired in the first place. This is because some clients can be difficult to work with—especially in production.

This is most often the case when the client does not understand the production process but makes requests that have a profound effect on it. For example, a client may decide on the second day of shooting that he doesn't like some elements of an actor's wardrobe, perhaps the fact that the employee is not wearing a company ID badge. To remedy this, he asks that the badge be included from this point forward. He doesn't realize that his request will destroy the program's continuity.

On the other hand, clients can also be quite helpful, sometimes catching mistakes before they are recorded or suggesting ways to streamline the production process as a result of their technical know-how on a particular subject. In dealing with both helpful and difficult clients, the corporate director should know and keep in mind certain "client handling" rules.

Clients always come first It is the client who is most likely paying the bill for the director's services and all or most other production costs. It is also the client who is responsible to a higher authority for the end product that investment produces. Most important, it is the client who, in the end, will be left to present the program to company employees.

All these elements add up to the fact that clients are very important persons in the director's life, and they should be treated as such. In fact, developing and maintaining solid client relations is one of the primary aspects of a corporate director's business.

Never play "Hollywood" Some directors, often the less experienced ones, feel they must project an important, Hollywood image for their clients. They assume this will make clients feel they are dealing with an influential, creative talent. The experienced corporate director realizes that this type of ego trip is not only unnecessary but often leads to poor working relationships and damaging consequences for the project.

This is because a Hollywood image is a signal to a client that a director is not approachable. This, in turn, reduces a client's comfort level and takes away his ability to contribute. It creates a barrier between these two key players in the production process. Discomfort and barriers are the exact *opposite* of what the director wants as he attempts to effectively produce a client's script.

By contrast, the director who projects an image of a mature, hardworking, down-to-earth businessperson has a much better chance of creating a comfortable atmosphere between him and his clients. This

You may find that your creative ideas have been nixed by some executive who, more likely than not, has no idea of what visual communications are really about. Use your negotiating skills to try to persuade the client to do it your way, but be prepared to lose the battle—gracefully.

—Donna Chandler
Director

Case Study 1 *The Hollywood Director*

I knew a director who was quite vain and very concerned with his personal image. He tended to push the limits of corporate acceptance with his style of dress and grooming. He also liked to impress his clients with the "mystique" of his "cinematic knowledge" and his status as a Hollywood director.

He would frequently drop comments like, "I have to be in Washington next week to shoot a crowd scene with 400 extras for a feature I'm working on," or "My agent has just lined me up some episodic television work, so I'll probably be tied up in meetings with the networks next week."

Regardless of his ego, he was an effective director who had a reputation for delivering quality work. As a result, he was employed frequently by corporate producers.

His welcome ran out at one company, however, when a client wanted several of the scenes reshot due to lighting she felt was overly dramatic for her program.

The director was so irritated by this "gall" on her part that he told her, in so many words, that she was hardly the person who should be criticizing an artistic subject like lighting. He let her know that he was an expert on visual and performance aesthetics and that his director of photography was one of the best in the state.

The client, unlike many others this director had worked with, was not intimidated. Best or not, she said, she felt the lighting was too harsh and she wanted the material reshot.

This led to a confrontation between her, the producer, and him. The producer had little choice; he sided with the client and insisted that the material be reshot.

The director finally consented to reshoot the scenes, but he let it be known that he didn't like his status as a director or his aesthetic judgment being called into question.

Following that project, he received no more offers from that producer, and as time went on, he experienced the same types of problems while working with other companies.

---

inevitably leads to the type of fertile environment in which the corporate production process can flourish.

Seriously consider client input When the client makes a suggestion in the middle of production, there is often a temptation to patronize her with a skillfully worded, evasive comment and move forward on the assumption that she doesn't really know what she's talking about. Many times, however, a client's input, though not educated in terms of production, can be exactly what's needed to clear up some bothersome issue.

I once directed a program in which a female host was cast as an operator supervisor. As we rehearsed the first scene in the studio, it became obvious to me that something wasn't quite right. She seemed to lack seriousness or authority. Meanwhile, the client looked on. Though she said nothing, she didn't seem particularly enthusiastic about the scene herself.

Finally, I left the control booth and went out onto the studio floor. As I was discussing line delivery with the actress, the assistant director's voice suddenly came over the stage PA system.

"Ray," she said, "The client has a theory."

"What's that?" I asked.

The client's voice then came over the PA. "I think it's her pants," she said.

"Her pants?" I asked.

"Right. Our supervisors usually wear dresses."

As I thought about it, I realized she was absolutely right. I had been focusing so intently on getting a *relaxed* performance that I had failed to realize the actress's wardrobe was *too* casual. We arranged for a quick change, and sure enough, with her new, businesslike and authoritative look, she became much more believable in the role.

Make clients feel they are a part of the process Finally, the experienced director knows that client approval is a must on his project, and he also knows that relationships in which he is seen as above reproach often lead to discontentment and friction. Clients are much more likely to approve, enjoy, and pitch in on a project if they feel they are on an equal level with the director and working together as a team toward the same goal—an excellent program.

The Skills

Having now gained a working knowledge of the corporate environment and its key players, you have put the first cornerstone of that foundation we mentioned earlier firmly in place. The second cornerstone involves learning the specific technical aspects of corporate television production. It is made up of four types of indispensable director's skills:

Production skills knowledge of the equipment, the process, and the people involved in the creation of video programs

Aesthetic skills knowledge of the script, human, technical, and general aesthetic qualities that a director applies constantly to guide his shots, scenes, and programs

Interpersonal skills the people skills required to supervise the crew, consult with the client, coach the actors, and provide a business perspective to the producer

Judgment skills the critical skills required to make valid creative and business decisions, at times under extreme pressure

PART II

Skills

Production Skills

4

❑ ❑ ❑ ❑ ❑

The director's production skills encompass a complete working knowledge of the entire production process. That knowledge includes an understanding of

Production equipment a close familiarity with all types of production and editing equipment

Production phases a thorough understanding of the four-part production process

Production people a solid understanding of what each crew member is responsible for on a production

A working knowledge of each of these areas allows the director to make critical judgments during the preparation, production, and editing phases of his program's development. For instance, in order to assign his assistant director the proper tasks during preproduction, the director must know specifically what must be accomplished during this phase of the process. In order to decide whether a long or a short focal length is appropriate in production, he must know what type of image each will produce on the screen and any complications or ramifications involved in his request. In order to be able to ask a director of photography to make certain changes in the lighting, he must know what those changes will require in terms of time and possible complications and how they will affect his shot.

As you might guess, volumes have been written on these subjects and we could never hope to cover each in significant depth in one chapter of a single book. We can, however, arrive at a clear initial understanding of them that will help clarify their relationship to the directorial process. As with that process itself, this understanding should simply be considered a framework that will serve as a basis for much more in-depth work and study.

Production Equipment

A basic video production requires the following equipment:

a *camera* to view and provide a visual perspective of a scene or event

a *microphone* to allow for the recording of the audio or sound portion of the scene or event (in cases where audio recording is required)

a videotape recorder (VTR) to record the camera's and microphone's output
lighting equipment to ensure that the scene has the proper exposure levels and visual aesthetic quality
videotape and in some cases *audiotape* on which to store the visual and audio recordings
accessories, such as dollies, tripods, cables, stands, batteries, and many other types of support and modification equipment

Cameras

There are three primary types of video cameras: camcorders, standard field cameras, and studio cameras.

Camcorders are normally used on location shoots. A camcorder has a VTR attached as a part of its body, as shown in Figure 4.1.

Standard field cameras are also typically used on location. They feed the visual signals seen by the camera's lens to a *separate* VTR. This is usually a portable cassette recording unit, such as the one in Figure 4.2.

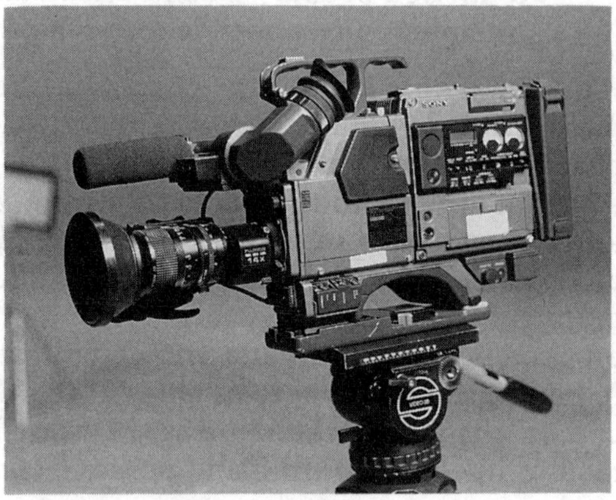

Figure 4.1 Sony Betacam is a widely used camcorder in both corporate and broadcast television. VTR is the upper-rear part of the unit. Attachment at end is battery compartment.

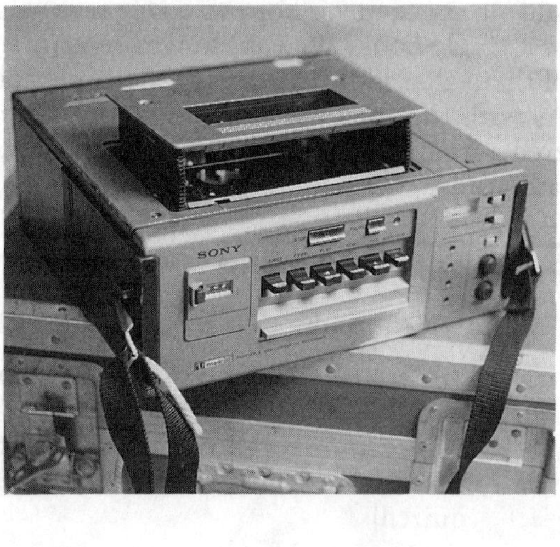

Figure 4.2 Until the advent of camcorders, portable ¾-inch VTRs such as this were the mainstay of corporate television. Many ¾-inch VTRs are still used.

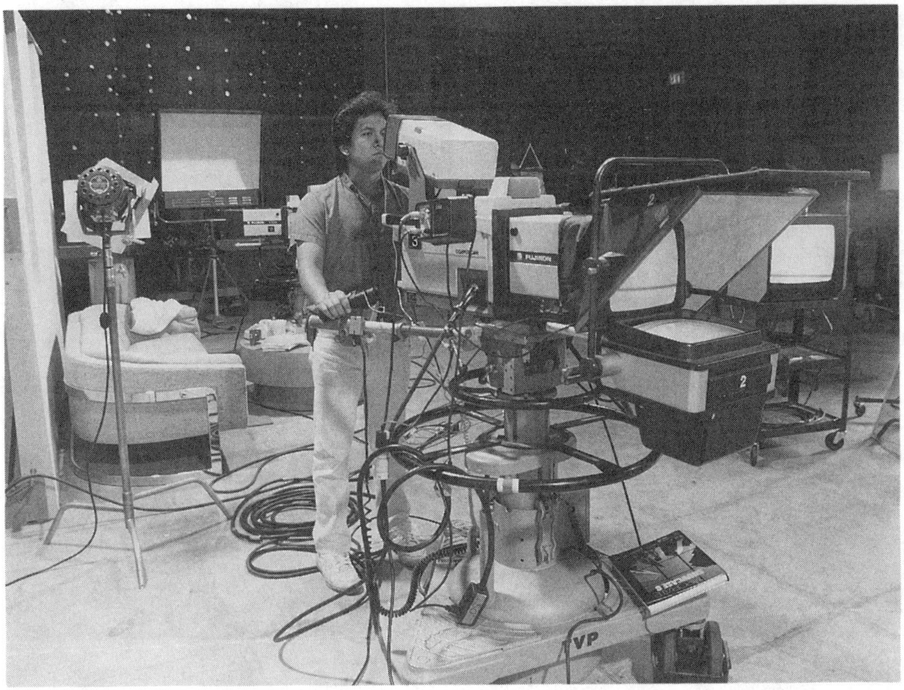

Figure 4.3 Hitachi studio camera in use.

Finally, studio cameras, as seen in Figure 4.3, often feed the video signal to a reel-to-reel VTR, as in Figure 4.4. Video cameras can also utilize each of the various types of VTRs. A reel-to-reel VTR, for instance, can be brought on location, or a cassette VTR can be used in the studio.

Whatever the situation, these cameras each manipulate a visual image in several common ways. They allow for exposure adjustments, color adjustments, image registration adjustments, and conversion of the visual image into an electronic signal that can be recorded on videotape.

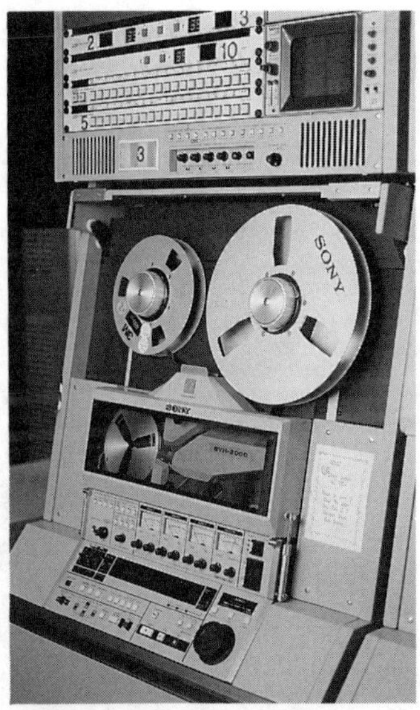

Figure 4.4 One inch reel-to-reel VTR used for studio production and editing applications.

Videotape Recorder (VTR)

Camcorder VTRs, such as the broadcast quality Sony Betacam, are small, lightweight units that record visuals on a high-speed, specialized system that produces very high-quality images. Although camcorder VTRs are normally attached or "docked" to the camera, most can also be disconnected, allowing the camera to feed its signal into an "outboard," undocked VTR, as in the arrangement displayed in Figure 4.5.

In corporate programming, this other recorder might be a U-matic, ¾-inch cassette VTR. Three-quarter-inch VTRs had been a mainstay of corporate video productions through the early 1980s, until the advent of camcorders. Though they may not produce what are considered broadcast-quality images, ¾-inch units are still used today in many corporate and even some broadcast productions.

Studio configurations recording to high-quality cassette and 1-inch reel-to-reel VTRs, *are* of broadcast quality and, in fact, are the type of

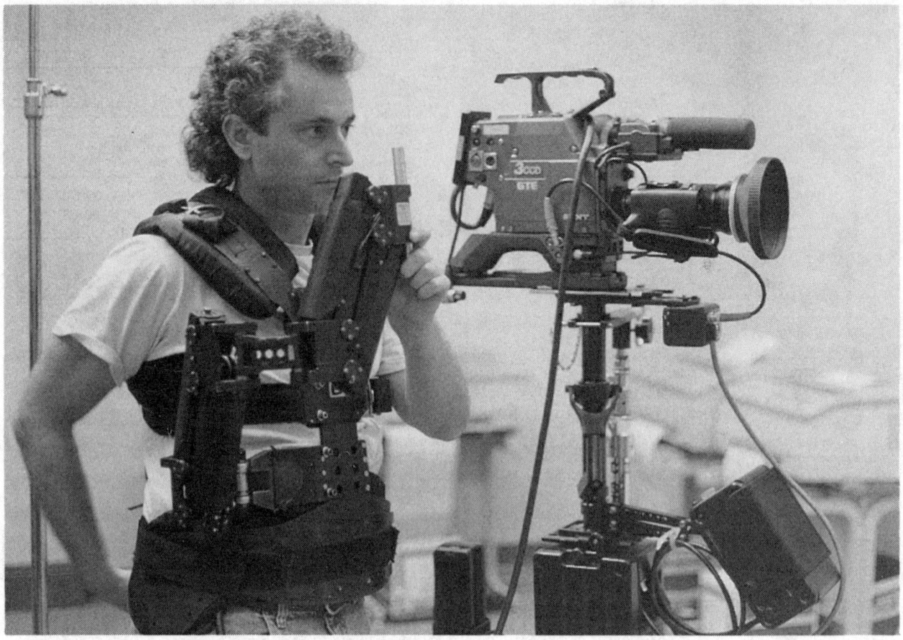

Figure 4.5 Sony Betacam feeding an *undocked* VTR. Note that only front half of camera appears to be on mount. VTR is out of sight, connected to thick cable.

equipment used by the major networks to produce their own video programming.

A corporate video department might use one or perhaps all three of these types of camera/VTR arrangements in its productions.

Microphones

Just as cameras allow the visual signal to be input, altered, and transported to the VTR, microphones allow the same process to take place with the audio signal. The three types of microphones (mics) most often used in video production are shotguns, lavalieres, and hand-held.

Shotgun mics Shotgun mics are long and slender like the barrel of a shotgun (Figure 4.6). They are normally placed near the sound source—

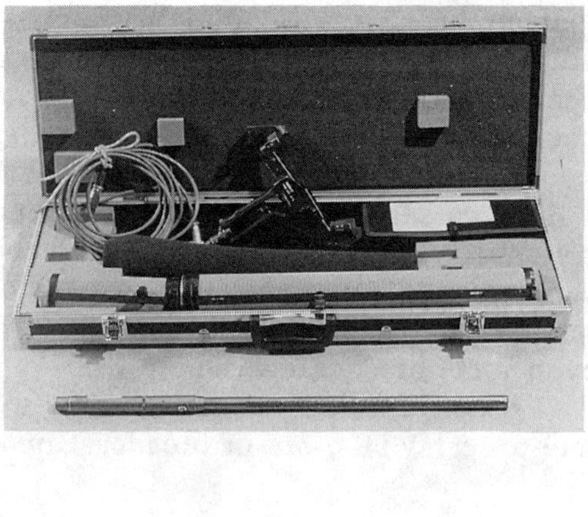

Figure 4.6 Shotgun microphone (accessories in case behind it) is used often in corporate TV, as well as in broadcast and feature film production.

usually the actor, employee, or executive—just out of the camera's frame. This is often accomplished with the use of a "boom" or "fishpole" held by the audio recordist's assistant or "boom operator" (Figure 4.7).

Audio signals are picked up by the mic, transported along an audio cable, and most times fed into a mixer, as in Figure 4.8, which provides basic volume and quality control. The signals then pass out of the mixer along another audio cable and into an input jack on the VTR. In the case of a camcorder, that audio input is on the camera itself (Figure 4.9).

Lavaliere mics A modern lavaliere mic (Figure 4.10) is a tiny capsule unit at the end of a very thin wire. On the opposite end of the wire is the unit's battery housing-cable connector. The capsules are clipped onto a tie or lapel or, when they must be hidden, taped under a seam or collar.

Though lavalieres react to audio signals somewhat differently from

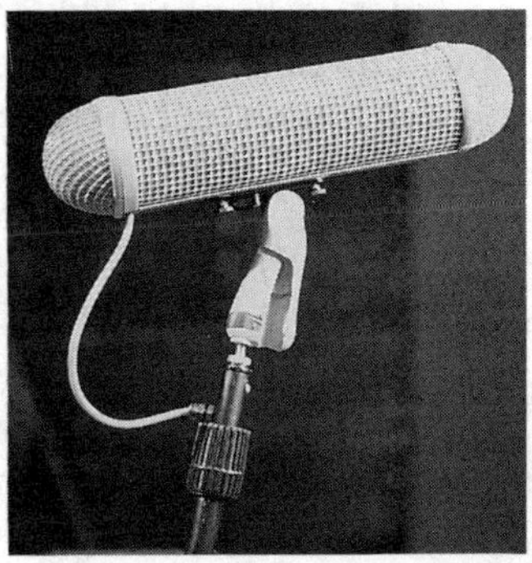

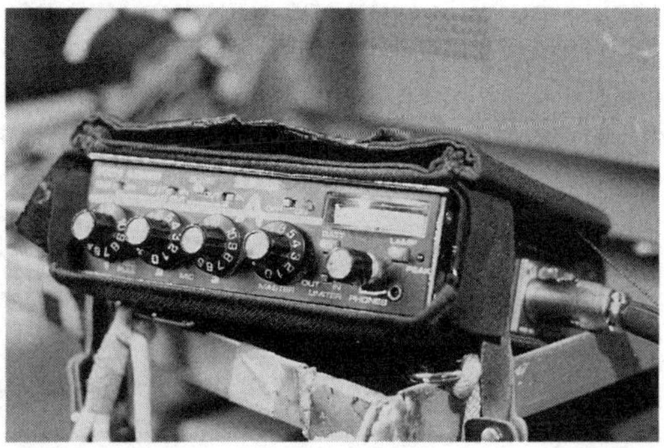

Figure 4.7 (left) End of "fishpole" or "boom" with a shotgun mic (inside zeppelin) attached.

Figure 4.8 (right) Small field mixer interfaces between microphone and the VTR. It offers several outputs and control functions.

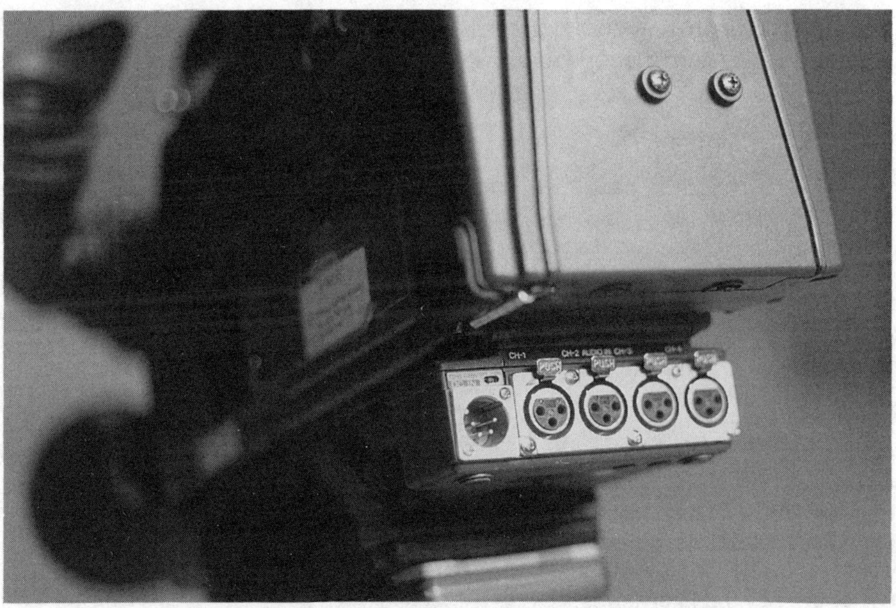

Figure 4.9 Audio inputs on back of Sony Betacam.

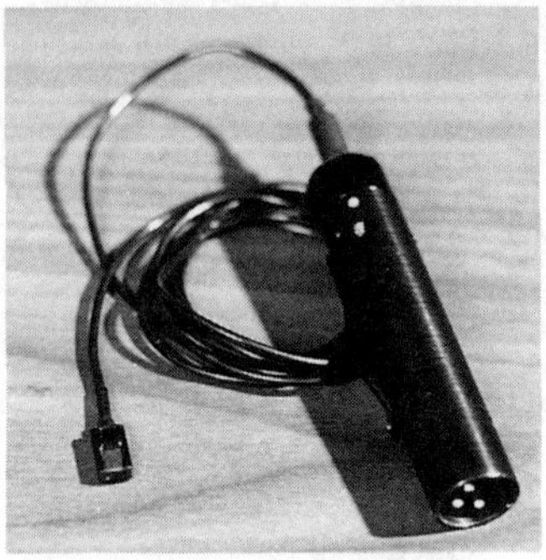

Figure 4.10 Lavaliere microphone. Like the shotgun, it is used frequently in both corporate and broadcast television productions. Mic element is small capsule at left. Barrel at right holds battery and allows for cable connection.

the way shotgun mics do, they send signals to the VTR in exactly the same way—along a cable and in most cases through a mixer.

Hand-held mics Hand-held mics are those used most often by news reporters or emcees (Figure 4.11). Though they are also used in corporate programming, they are much less frequently found on shoots than shotguns or lavalieres.

Wireless mic systems Wireless mic systems convert the standard audio signal to a radio frequency (RF) signal and send it from a transmitter located with the mic to a receiver unit, usually located with or somewhere near the VTR (Figure 4.12). Wireless systems are used when audio cables will be unsatisfactory for the shot being recorded. All three types of mics just discussed can be used on wireless systems.

Lighting Equipment

Lights provide both the illumination needed for a video system to record the image and the aesthetic quality required to help make the scene convincing or appropriate to the mood the director has requested. In order to understand how lights do this, we must first touch briefly on the subjects of luminance, chroma, and color temperature.

Luminance Light intensity is called luminance by video engineers. Luminance can be easily measured on a waveform monitor, such as the one shown in Figure 4.13.

If there is too much luminance in a video shot, it is said to be too "hot." If there is insufficient light, it is said to be "down" or "underexposed." One primary purpose of production lights is to provide the optimum light quantity or luminance "levels."

Chroma Chroma is the engineering term for color. It is measured and adjusted on a vector scope, as in Figure 4.14. A second function of lights

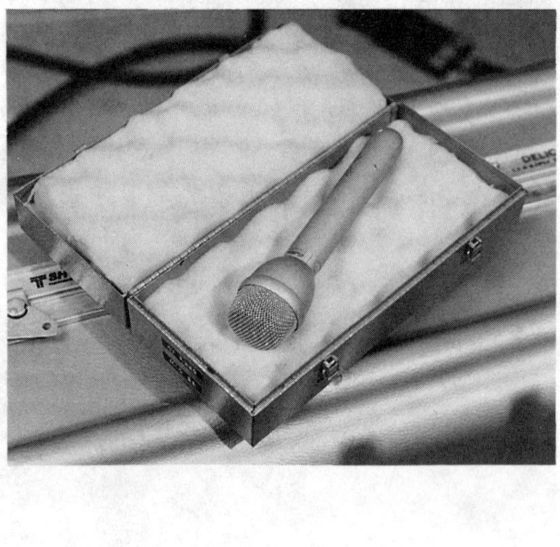

Figure 4.11 Hand-held mic, often used for news-style interviews. It may be held by an interviewer or by the interviewee.

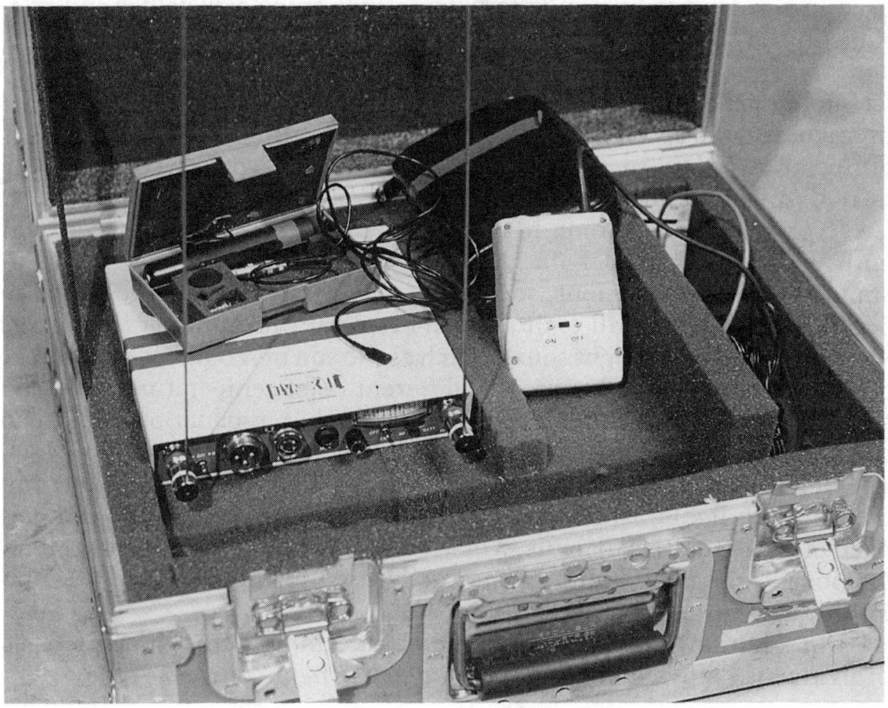

Figure 4.12 Wireless microphone system. Mic (in box) and receiver unit is white rectangular unit on left, transmitter unit is on right. Output of the receiver is fed to VTR.

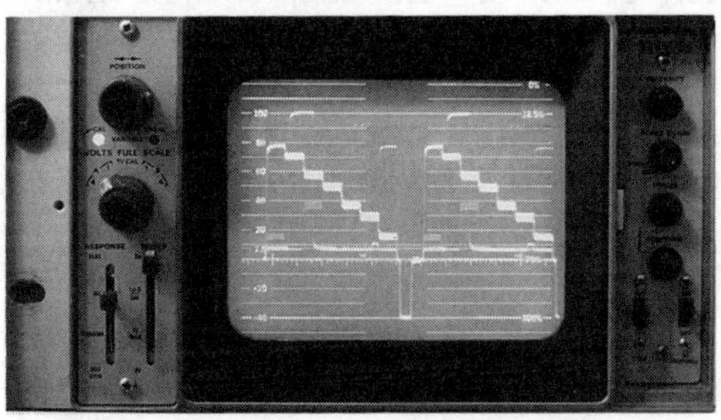

Figure 4.13 Waveform monitor, used for measuring light intensity and other factors. Reading shown is for color bars set up. Highest "steps" are brightest light sources.

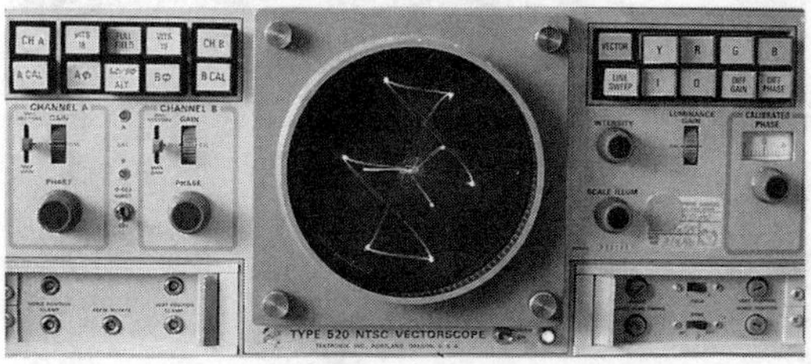

Figure 4.14 Vector scope measures and allows for adjustment of color phases of video signal. Point locations represent different color values.

used in a video production is to help provide true-color values and relationships or, if necessary, to alter the color quality.

Color temperature Color temperature is the industry standard measurement used to specify the color value of light falling on an object. Color temperature refers to the color of light emitted by a carbon filament heated to various temperatures on the Kelvin scale.

For instance, a carbon filament heated to 3200 degrees kelvin will emit an orange hue of light that mixes naturally with light found in standard *interior* situations, such as homes. A carbon filament heated to 5600 degrees kelvin will emit a much more pale, bluish light that mixes well with an *exterior* light source, such as the sun or moon. Since different types of production lights provide different color temperatures, you can see that besides providing illumination for a shot, production lights also color it to some degree.

The two most common types of lights used in corporate production are tungsten lights, which provide light at 3200 degrees kelvin, and halogen metal iodide (HMI) lights, which provide light at 5600 degrees kelvin. Obviously, tungsten lights are normally used for interior shooting and HMI lights for exterior shooting.

In addition to illumination and color temperature, different types of lights also provide either "hard" or "soft" light sources.

Hard lights Hard lights throw a direct beam on the subject. They create definite or "hard" shadows on an actor or object. They are lights such as the open-faced units seen in Figure 4.15, and those with a glass lens over their openings, known as Fresnels (Figure 4.16). Hard lights can be "spotted," which means narrowing their beam, or "flooded," which means broadening their beam.

Soft lights In soft lights, the beam is much broader and more diffused since it is "bounced" out of a reflective hood. This results in softer shadows on the actors or objects (Figure 4.17).

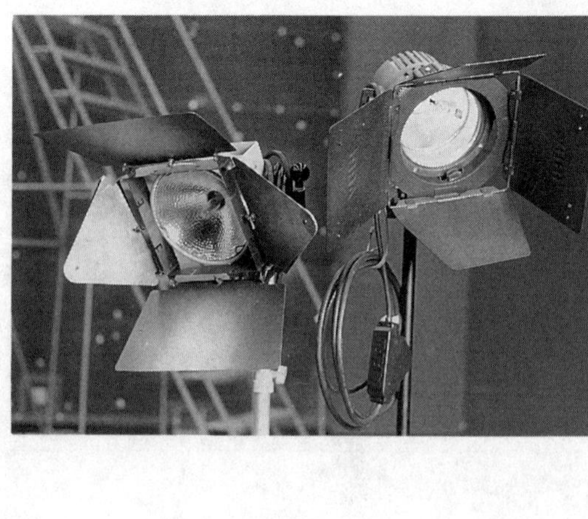

Figure 4.15 Open-faced lights, such as Lowell (left) and Arri (right), are used frequently in corporate television.

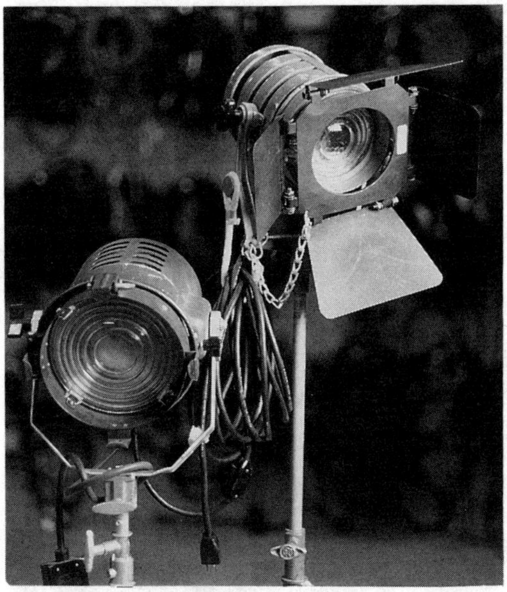

Figure 4.16 Fresnel lights, such as Colortran (left) and Mole Richardson (right), have glass lenses over the bulbs. This slightly diffuses the light source.

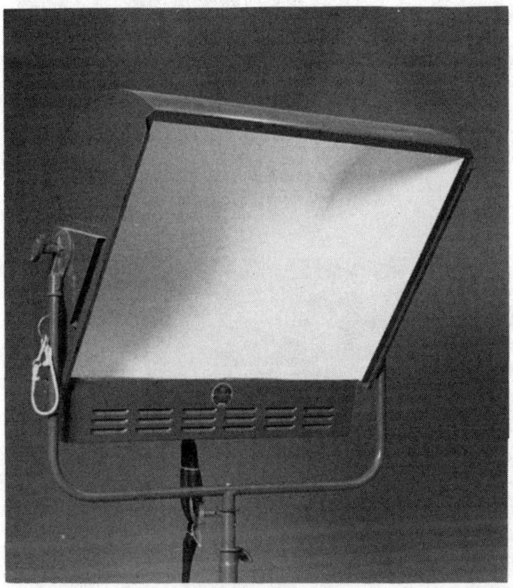

Figure 4.17 Soft light bounces light from lower section out of a reflective hood. Its light is thus softer, broader, and more "ambient" in nature.

Videotape and Audiotape

Videotape receives the video and audio signals and allows them to be recorded for later playback or duplication. Beta cassette videotapes are used in camcorders, such as Betacams; ¾-inch U-Matic cassettes are used in ¾-inch systems; and 1-inch reels are used in studio VTRs (Figure 4.18).

One-quarter-inch is normally used in corporate "audio only" recordings. It too can come in either cassette or reel-to-reel formats.

Accessories

Production accessories range from gaffer's tape (which is used for everything from covering hot spots to sticking pictures on walls) (Figure 4.19)

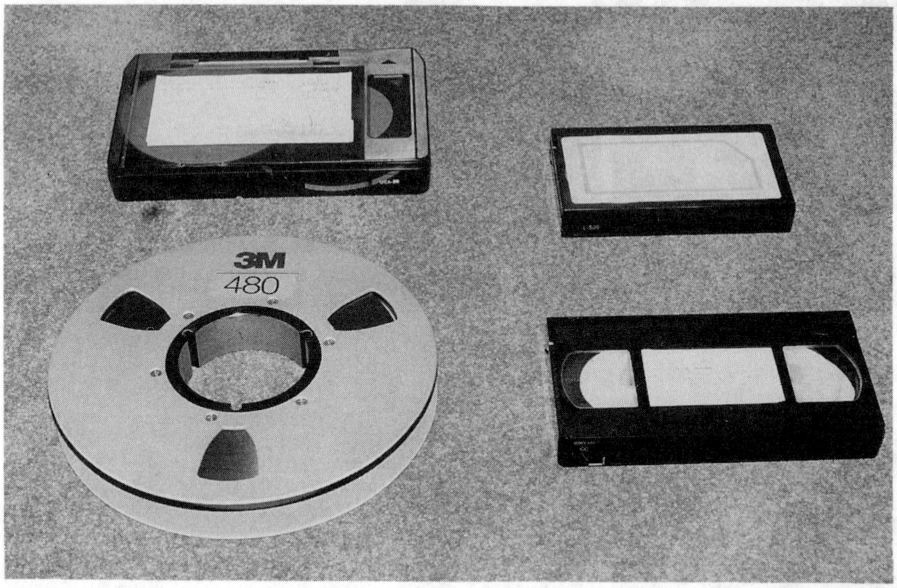

Figure 4.18 Videotape. Clockwise, starting at lower left: 1-inch, reel-to-reel; ¾-inch U-Matic cassette; Beta cassette; VHS cassette.

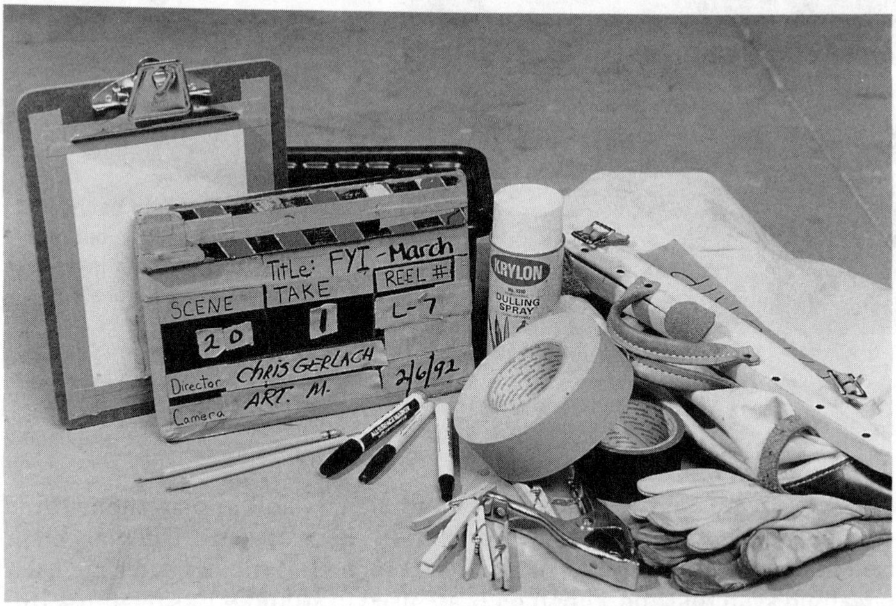

Figure 4.19 Gaffer's tape, slate, black markers, and other production accessories.

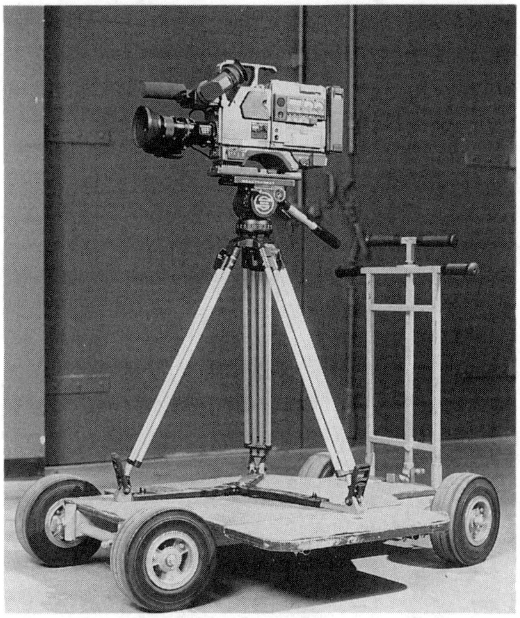

Figure 4.20 Dolly places camera on wheels, allowing it to move with a very smooth motion.

to cranes, dollies, and jib arms, which are used to place or move the camera according to the director's instructions (Figure 4.20).

Lighting accessories include "C" stands (Figure 4.21), flags and scrims (Figure 4.22), silks and nets (Figure 4.23), "bounce" cards and colored gels (Figure 4.24), and an assortment of special clips, arms, and fixtures (Figure 4.25).

Camera accessories encompass various lenses, filters, zoom motors, handles, head covers, and other attachments.

Sound accessories include zeppelins, which cut down on wind or extraneous noise, and wind socks, which slip over zeppelins to increase their effectiveness. Clips are used for attachment to clothes, and assorted types of handles and head covers come in handy in various situations.

Among additional accessories are batteries and power supplies for cameras, mics, recorders, and some lights; audio, video, and electric cords; generators, when power must be supplied; slates for recording scene information; and other items, such as makeup kits for actors and dulling spray to reduce glare.

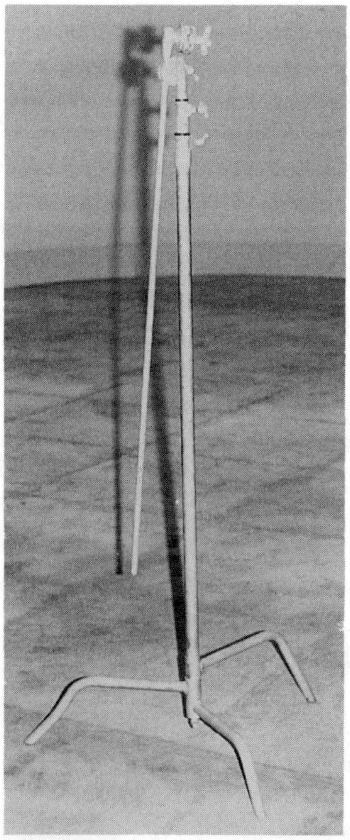

Figure 4.21 "C" stand, the lighting workhorse for the industry.

Production Phases

Television and film production usually takes place in four distinct phases: *writing, preproduction, production,* and *postproduction.* As with equipment, considerable time and on-the-job experience are required to become completely knowledgeable about each. What follows, however, is a brief overview for purposes of familiarization.

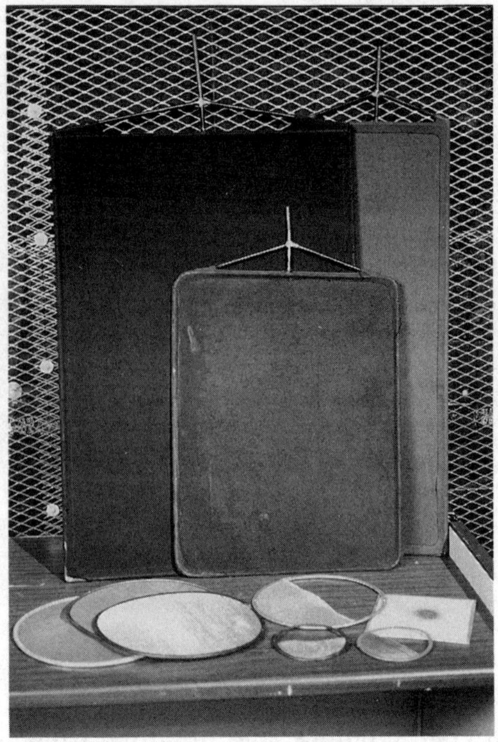

Figure 4.22 Flags standing and scrims on tabletop. Both are used to control lighting. Flags cut light off, and scrims reduce its intensity.

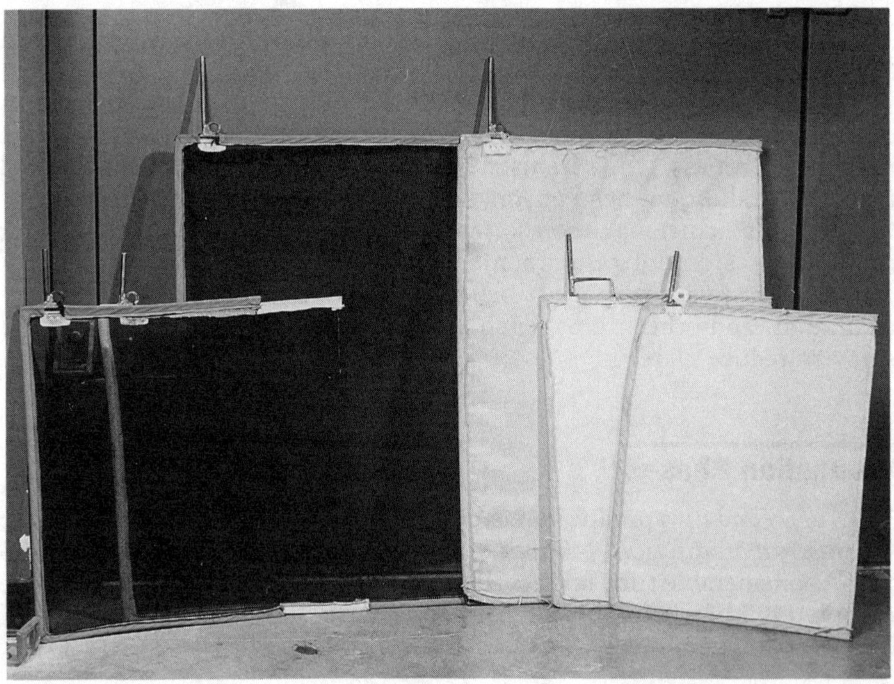

Figure 4.23 Nets on left and silks on right. Silks diffuse light. Nets act like scrims, reducing its intensity.

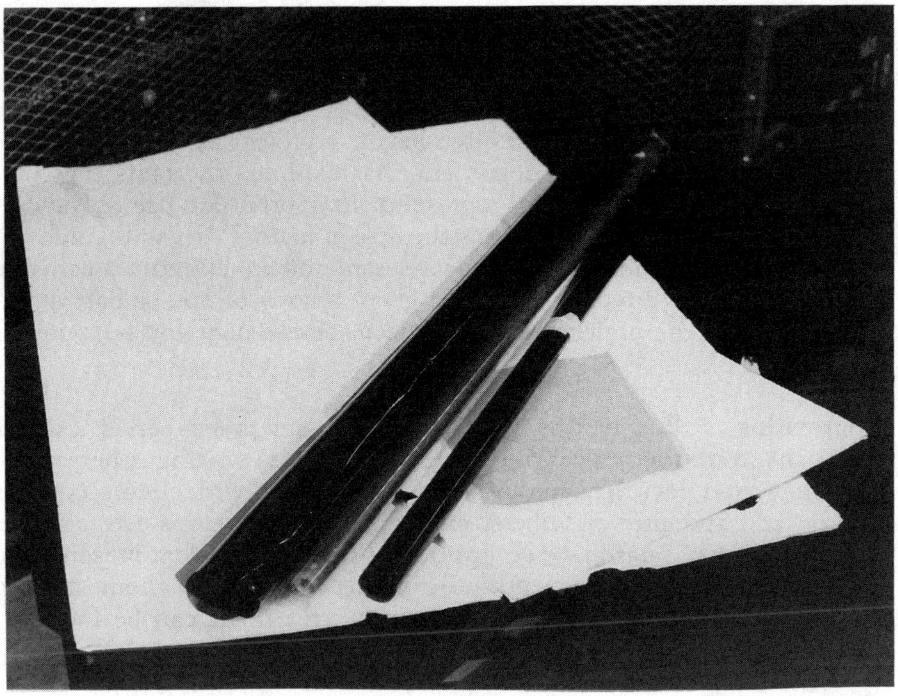

Figure 4.24 Bounce cards (white foamcore material) and rolls of gel material. Cards reflect light; gels color it.

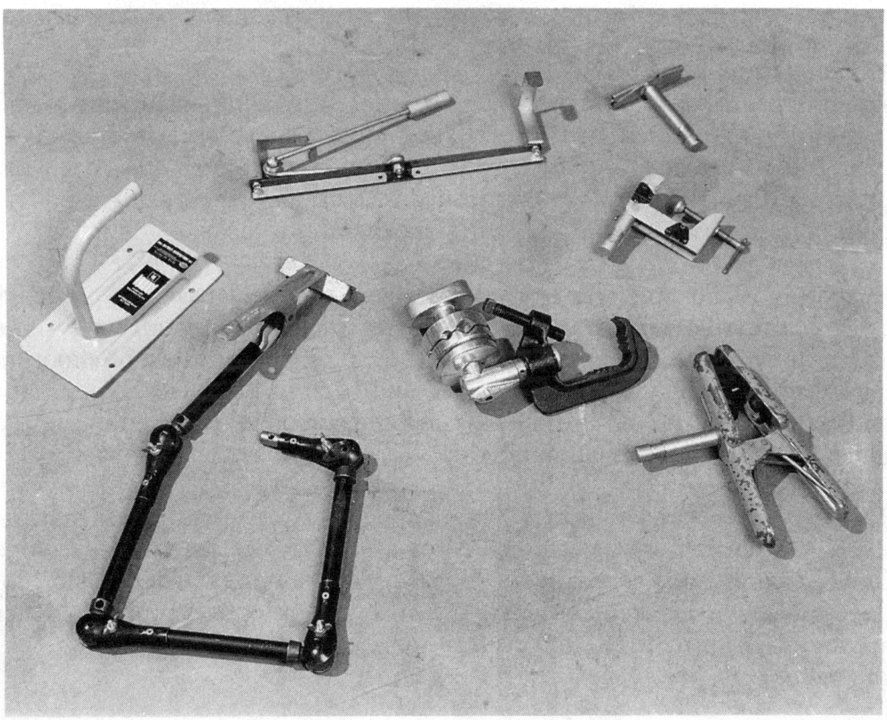

Figure 4.25 A variety of clips, arms, and fixtures used in production lighting.

Remember who your audience is. The issues in your program must appear to be important to your viewer as well as your audience.

—Dutch Bickley
Executive Producer

Writing

When a client comes to the corporate producer with a communication problem, a corporate video project is in its inception.

Program design The producer often begins a project by writing a needs assessment. This is a brief document that explores the facts that will determine whether the project represents an appropriate use of the corporation's money. It also examines the design factors that will guide the project's entire development. Needs assessments are extremely valuable documents, and a director should be given a copy of one as part of his introduction to the project. A sample needs assessment can be found in Appendix 1.

Scriptwriting Following this initial assessment and design period, a writer is brought in on the project, unless the producer is writing it herself.

The writer does the appropriate research; meets with clients, content experts, and audience members; and, from this, develops a treatment. The treatment is a narrative description of the program that presents the proposed story line or concept and explains to both the client and the producer the basic visualization. A sample treatment can be found in Appendix 2.

After approval of the treatment, the writer develops the script. This is usually accomplished in approximately three drafts, each of which are presented to the client and producer for their input and eventual approval. For a sample script, see Appendix 3.

The writing phase of production ends with the writer's delivery of a client- and producer-approved shooting script.

Preproduction

Preproduction is the process of gathering, organizing, confirming, and scheduling all elements required to complete a successful production. It is at the outset of preproduction that the director is brought in on the project. Although a great many minor tasks take place during a typical preproduction period, the following are the major tasks.

Budgeting The producer typically budgets a project based on a breakdown of the shooting script. Her breakdown takes into account such items as the number of actors needed and their fees, the number of shooting days required, the crew size required and each of their fees, equipment and facility rentals, and the editing time and personnel required. A sample budget can be found in Appendix 4.

Talent auditions and bookings Normally, the director and an assistant—usually a production assistant (PA) or assistant director (AD)—arrange for actor auditions, selections, callbacks, and bookings. Clients and producers frequently become a part of the talent selection process, either approving the director's choices or in some cases actually making choices themselves.

Crew bookings Crews are usually hired by the PA or AD based on the producer's or director's request. Typically, the director will be most concerned with his director of photography (DP) or videographer, because this is the person who will execute the lighting and major camera work. The crew sheet in Figure 4.26 lists job positions typical of a corporate shoot.

Equipment reservations or rentals The AD or PA books the camera, lights, mics, and other equipment either by reserving in-house equipment owned by the corporation or by calling equipment rental houses. In the

PRODUCTION CREW ASSIGNMENTS

DISTRIBUTION:
White – Production Coordinator
Yellow – Producer / Show File
Pink – Director
Goldenrod – Ad

TITLE IN GOOD COMPANY V# 88W1270

PRODUCER Ray DiZazzo DIRECTOR Ray DiZazzo AD Cindy Wermers

	DATES	CALL TIME	LOCATION &/OR CONTACT NUMBER
PRE-PRODUCTION	10/23 - 11/6/89	8:00 a.m.	5710 or 5719
PRE-PRODUCTION MEETING	11/6/89	12:30 p.m.	5710 or 5719
PRE-LOAD			
BUILD & LIGHT			
TECH			
PRODUCTION REMOTE	11/7/89 - 11/16/89	7:00 a.m.	See attached schedule
STUDIO			

STUDIO		**REMOTE**	
TD / VTR / VC / TITLE		E.I.C.	
AD / SCRIPT		AD / SCRIPT	Cindy Wermers/Remy Mackenzie
LIGHTING DIRECTOR		D.P.	F. Martin
FLOOR MANAGER		AUDIO	Eric Pierce
CAMERA OPERATOR #1		TELEPROMPTER / CUE CARDS	Mac Daigre
CAMERA OPERATOR #2		GAFFER	Rafael Sanchez
CAMERA OPERATOR #3		GRIP	Rudy Rosales
AUDIO		Grip	Henry Coccetti
GAFFER		OTHER	Steve Lippert (Props)
GRIP			Judy Yonemoto (Make-up)
TELEPROMPTER			Alissa Williams (PA)
OTHER			John Lifavi (boom op)
			Jeff Hall (grip)

SPECIAL REMARKS We will also be using the announce booth and control room on 11/6/89 for a v.o. session.

Figure 4.26 A typical crew sheet used to make all involved in the production aware of dates, call times, and other basic information.

case of rental houses, the AD arranges to have the equipment delivered the day before the production and picked up the day after.

Studio and editing facilities are reserved or rented in much the same way. A typical facility/equipment reservation form used by one large company can be seen in Figure 4.27.

Location and set arrangements On feature films and network television programs, professional set designers and builders create sets under the supervision of an art director, who in turn reports to the director.

In corporate program production, the director often designs his own, much less elaborate set, and supervises having it built and lit. The DP and two or three crew members, perhaps the gaffer and two grips, may

SCHEDULE REQUEST/PLANNING FORM Rev 11/4/91

PRODUCER/CLIENT: PROJ # DATE:
PROJ/SHOW TITLE: NEW|__| CHANGE|__| ADDITION|__| CANCEL|__|
PHONE NUMBER: ADDRESS
SPECIAL NOTES:

PRODUCTION FACILITIES/SERVICES

STAGE SHOOT ACTIVITY	DATES BEGIN – END	TIMES IN – OUT	FACILITIES/EQUIPMENT INCLUDED SELECT ADDED ITEMS FROM OPTIONAL EQUIP
BUILD OR STRIKE	–	–	STAGE AREA
SET LIGHTING CHECK	–	–	STAGE/CAMERA #3/MONITOR (No Master Control Rm)
1 CAMERA SHOOT	–	–	STAGE/CAMERA #1/1" VTR #6/MASTER CONTOL RM
2 CAMERA SHOOT	–	–	ABOVE + CAMERA #3/GREEN & DRESSING RMS
PRODUCTION AUDIO	–	–	NARRATION CONTROL BOOTH

REMOTE SHOOT	BEGIN – END	IN – OUT	REMOTE SHOOT	BEGIN – END	IN – OUT
__CAMERA – Chip	–	–	__AUDIO KIT	–	–
__CAMERA – Tube	–	–	__LIGHTING KIT	–	–
__CRASH CART	–	–	__DOORWAY DOLLY KIT	–	–

NOTE: DETAIL REMOTE LIST MUST ALSO BE COMPLETED AND SUBMITTED TO ENGINEERING

POST PRODUCTION FACILITIES/SERVICES

OFF LINE – Days	BEGIN – END	IN – OUT	OFF LINE – Nights	BEGIN – END	IN – OUT
__OFF LINE A – DAY	–	–	__OFF LINE A – NIGHT	–	–
__OFF LINE B – DAY	–	–	__OFF LINE B – NIGHT	–	–
OFF LINE D – DAY	–	–	OFF LINE D – NIGHT	–	–

ON LINE (With Editor)	BEGIN – END	IN – OUT	NORMAL SET–UP + MASTERING SHOWN BELOW
__EDIT 1 – DAY	–	–	__KEY CHANNEL REQUIRED (Available In Edit 1 Only)
EDIT 1 – NIGHT	–	–	__D2 MASTERING – Added Cost | SELECT ADDED
EDIT 2 – DAY	–	–	__1" MASTERING – Added Cost |EQUIPMENT
EDIT 2 – NIGHT	–	–	Beta–SP MASTERING – Added Cost |FROM BELOW

DUPING SERVICE	ORIG AVAIL	DUPE DUE	DUPING SERVICE	ORIG AVAIL	DUPE DUE
FROM TO			FROM TO		

COMPUTER GRAPHIC	BEGIN – END	IN – OUT	SELECT OPTIONAL EQUIPMENT FROM BELOW
__BASIC/With Artist	–	–	AURORA COMPUTER AND STANDARD COPY CAMERA
Render/Xfer To Tape	–	–	SELECT VTR FROM OPTIONAL EQUIP/ADD COST

POST AUDIO	BEGIN – END	IN – OUT	SELECT OPTIONAL EQUIPMENT FROM BELOW
__SWEETENING	–	–	MASTER MIX SUITE ONLY
LayDown/Back + VTR	–	–	SELECT VTR/ADD COST (Master Cntl Rm/Edit Bay EXTRA)

OPTIONAL EQUIPMENT/FACILITIES – PRODUCTION & POST PRODUCTION

ITEMS	BEGIN – END	IN – OUT	ITEMS	BEGIN – END	IN – OUT
__ABEKAS DVE (Chnl A)	–	–	__ADDED CAMERA	–	–
__ABEKAS DVE (Chnl B)	–	–	__1" VTR (2000 or 2500)	–	–
__STAGE PROMPTER	–	–	__D2 VTR	–	–
__ULTIMATTE	–	–	__Beta–SP VTR (Standard)	–	–
__Title Generator (Dubner)	–	–	__Beta–SP VTR (Dyn Trkg)	–	–
__STILL STORE (Abekas)	–	–	__3/4 VTR (Pro BVU800)	–	–
__FRAME STORE (Adda)	–	–	__3/4 VTR (Ind VO5850)	–	–
__QUAD SPLIT	–	–	__Edit/Slide Copy Camera	–	–
__DRESSING ROOM	–	–	__SCREENING ROOM	–	–
GREEN ROOM	–	–	__BLUE ROOM	–	–
VAN (Personnel)	–	–	VAN EFP (Lift Gate)	–	–

ADDED COMMENTS

Figure 4.27 A facility/equipment reservation form used by one large corporate television department.

be brought in 1 or 2 days before the production to accomplish this (Figure 4.28).

On a feature film, initial location scouting is usually handled by a location manager and his staff. But again, in the corporate world the director is on his own in most instances. He must find locations that meet his needs and, working through his AD or PA, arrange to reserve, pay for, and schedule their use.

Many times in corporate programs, locations consist of company office areas and crew members' or clients' homes. When securing public locations, however, the AD or PA must apply for permits and insurance and pay the appropriate fees charged by cities, businesses, or local government agencies.

Blocking/storyboards/diagrams Having decided on his locations and carefully visualized the script, the director decides where his basic camera positions will be, the type of look he is after, and how the actors will move in and out of each scene.

This movement of actors in relation to the camera is called blocking. If the scene involves a host on camera, the blocking may be as simple as an entrance and an exit. If the scene involves characters in a dramatic situation, blocking becomes more complicated, however, and is usually finalized either in rehearsal or on location when actually working with the actors. Directors often draw simple storyboards and blocking diagrams, like those in Figures 4.29 and 4.30.

Rehearsals/meetings Talent rehearsals are the exception rather than the rule in corporate television. This is the case for two reasons. First, cor-

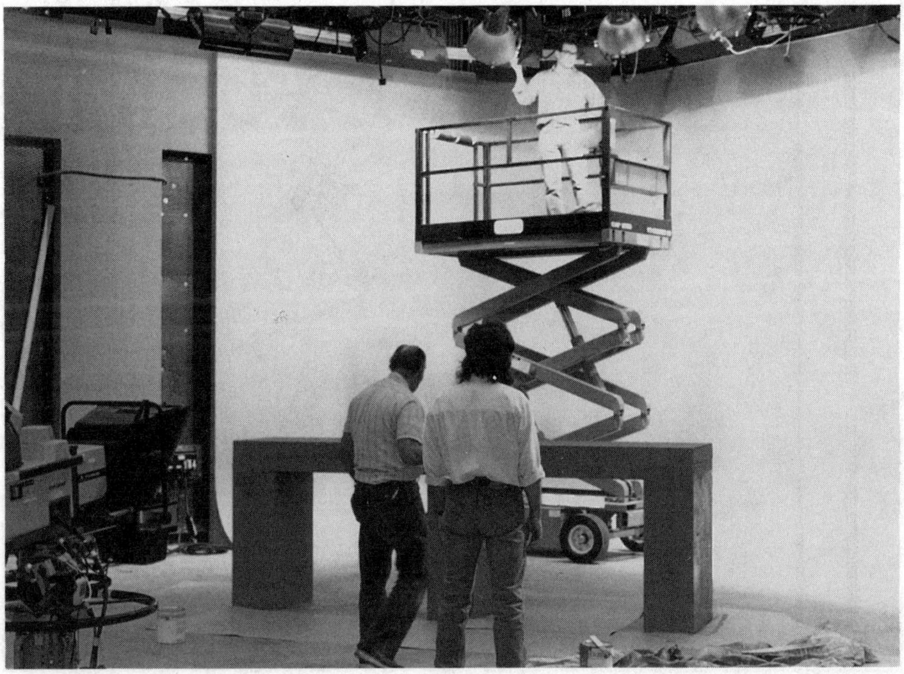

Figure 4.28 Crew at work in the first phases of set construction and lighting.

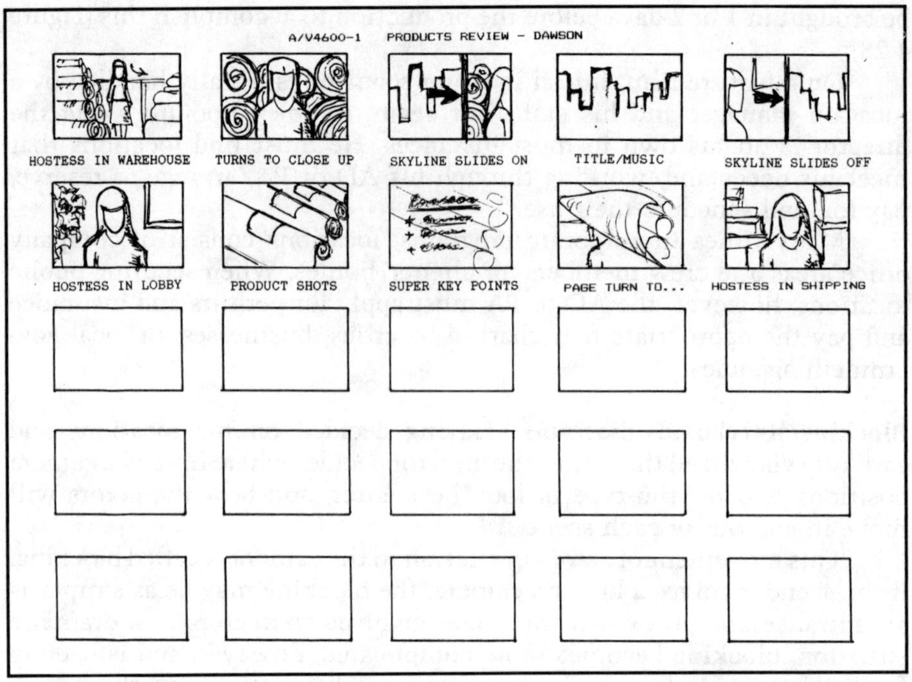

Figure 4.29 Director's storyboard, used to help develop blocking, determine image sizes, and edit visualization.

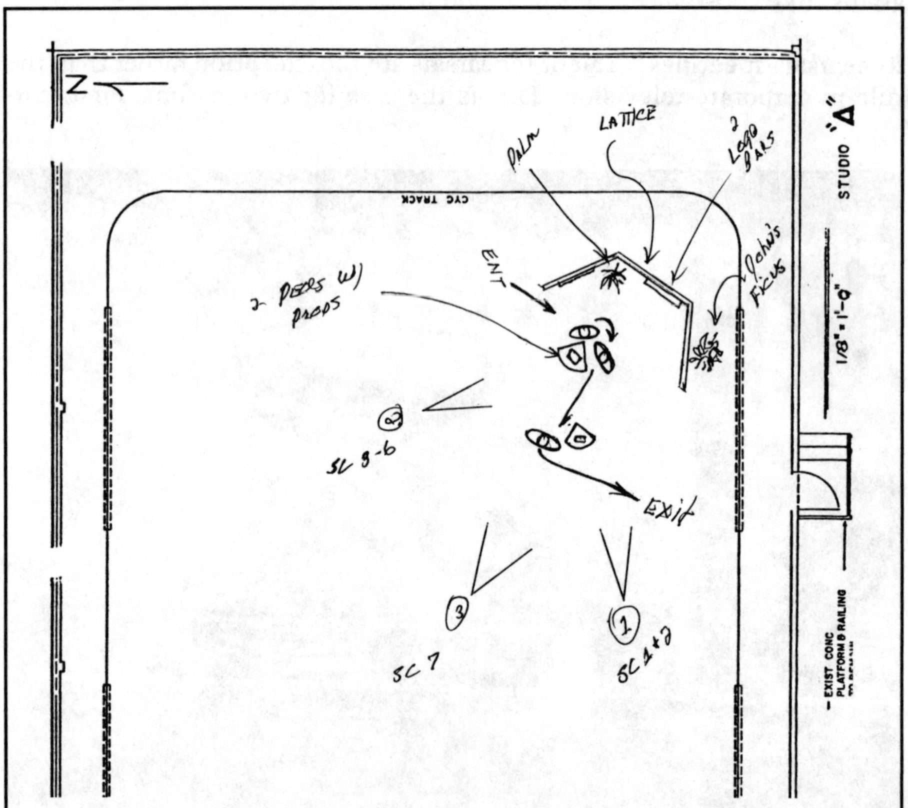

Figure 4.30 Director's blocking diagram. Used to establish camera positions and coverage.

porate budgets often will not allow for the extra expense. Second, most corporate scripts are considerably less involved than those for feature films or network television programs. When a dramatic or perhaps humorous program is being produced, however, it can actually save time and money to have the actors in for a rehearsal the day before the shoot.

Technical rehearsals are held more often in corporate production. A "tech," as it's called, is usually scheduled when a set has been built and multiple cameras will be used. In a tech, all crew members attend and assume their positions, and a stand-in actor may walk through the various angles, ensuring that the lighting and sound setups will work error free on the following production day.

Crew meetings are sometimes held during preproduction, when the shoot is a complicated one requiring updates by the various crew members. In the majority of cases, however, the director and his AD have most preproduction meetings, along with a brief meeting for the entire cast and crew, on the first day of production.

Production scheduling Finally, a schedule must be created to ensure that the various scenes are recorded in the most economical order, considering such factors as actors' daily fees, location fees and availability, travel times, equipment rental fees, and studio fees and availability. The director and the AD usually create the shooting schedule (Figure 4.31) after the director has made all location, talent, and basic blocking decisions. It is only with these elements decided that the director can accurately gauge personnel and logistic considerations, thus how much *time* each shot will take.

Production

Production is the shoot itself. It might take place at a location like a company cafeteria, on a prebuilt set, in an audio recording booth, or at a combination of these. Wherever it takes place, production is the short critical period of time when everything must work right.

Complicated human and technical elements are brought together in each setup to create an illusion of reality. If everything has been prepared properly, that illusion can be successfully created in the time allotted for each scene to be shot. If preparations have not been adequate, however, production can turn into a frantic catch-up game involving long, grueling days and continual compromises on picture, sound, and performance quality.

Location production Location production is usually carried out with a single camera. The crew drives to the scheduled location, and the lights, camera, and all accessory equipment are set up. The scene is rehearsed until the director feels it is ready to be recorded. At this point, a first master scene is shot. Typically, anywhere from two to ten takes of this scene are required to get what the director terms a "buy"—the actual take to be used in the program. Once the director declares the scene a buy, the AD circles that take in her notes on the master script and tape log for later reference by the editor.

Additional coverage of the same scene—or parts of it—is then re-

SHOOTING SCHEDULE - "IN GOOD COMPANY" DATE 11/13 CALL 7:00

PAGE	SCN	DESCRIP/COVERAGE	D'TM	TIME
NOTE:		CREW SETS UP PARTY AREA WHILE WE WORK FRONT DOOR.		
33	35	Mel comes to front door 2 angles: cam & past cam	8:30	1.5
26	25/26	Cindy enters lrm - A/fnt door sees empty living room.	9:30	1.0
4/5	5	Cindy Answers door. Geoff & Mon. enter/exit. - Dolly people pass camera.	10:30	0.5
4/5	5	As above from loft.	11:00	0.5
LUNCH 11:00				
6/7	6/7	Geoff moves to hors d'oeuvres table, talks w/Harold - dolly	2:30	2.5
7	8	Harold sees woman/man - POV/OS points out to Geoff	3:30	1.0
--	--	Mel's POV - lvng room. people clap. (For scn 35)	4:00	0.5
--	--	Cindy's POV - lvng room - empty (for scn 25)	4:30	0.5

CAST: Geoff, Monica, Harold, Cindy, Georganne, Barry, & Mel. 10 extras.

PROPS: food, (casseroles, salads) paper plates and utensils, drinks, Harold's glasses.

CREW: Everyone

CALL TIME AT VISNET: 7:00 a.m. crew
7:30 a.m. cast
10:00 a.m. extras (P.A. will P/U.)

SHOOT LOCATION: 5809 Woodglen Drive

Figure 4.31 Typical corporate television shooting schedule.

corded. This is accomplished by moving the camera to new positions, changing focal lengths, and recording the same action and dialogue or portions of it. Just as with the master scene, multiple takes are usually required at each position to finally record the buys.

During this entire time, the client or a technical expert may be present on the set, giving input on the technical accuracy of each shot. The producer may also be present, sharing an additional aesthetic perspective.

When the director is satisfied he has got the best performances and adequate coverage of the scene, the setup is "struck" or dismantled, and the crew, cast, and clients move to a new location to repeat the process.

Some setups are simple, involving only a camera and no lights (Figure 4.32). Other setups are complex, involving extensive lighting, camera movement on cranes or dollies, and complicated blocking of actors. Figure 4.33 shows a steadicam rig in use on such a shoot.

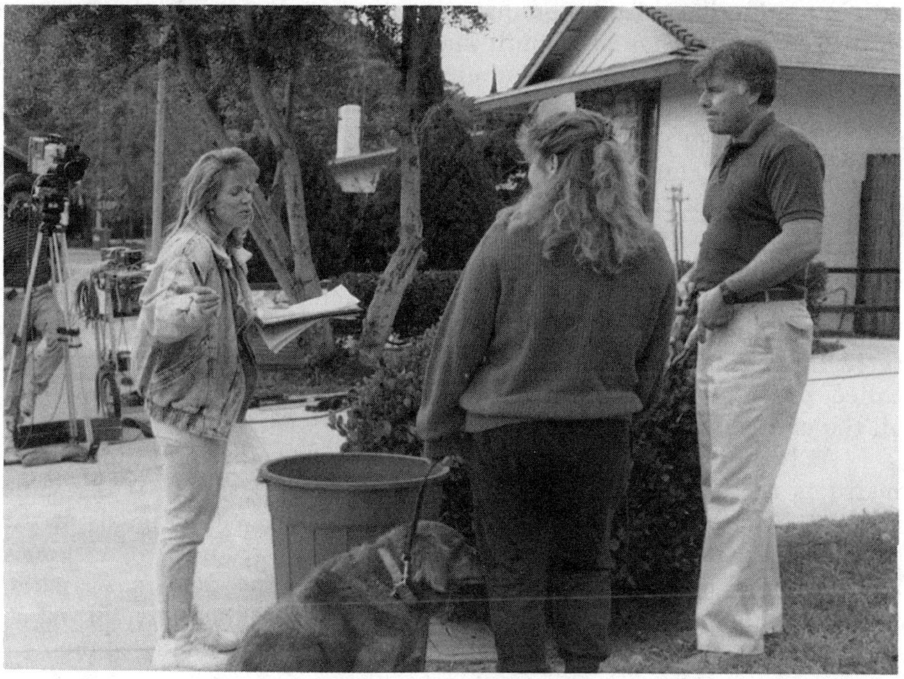

Figure 4.32 A simple location setup. Director, at left, works with talent, including dog. Camera is at far left in background.

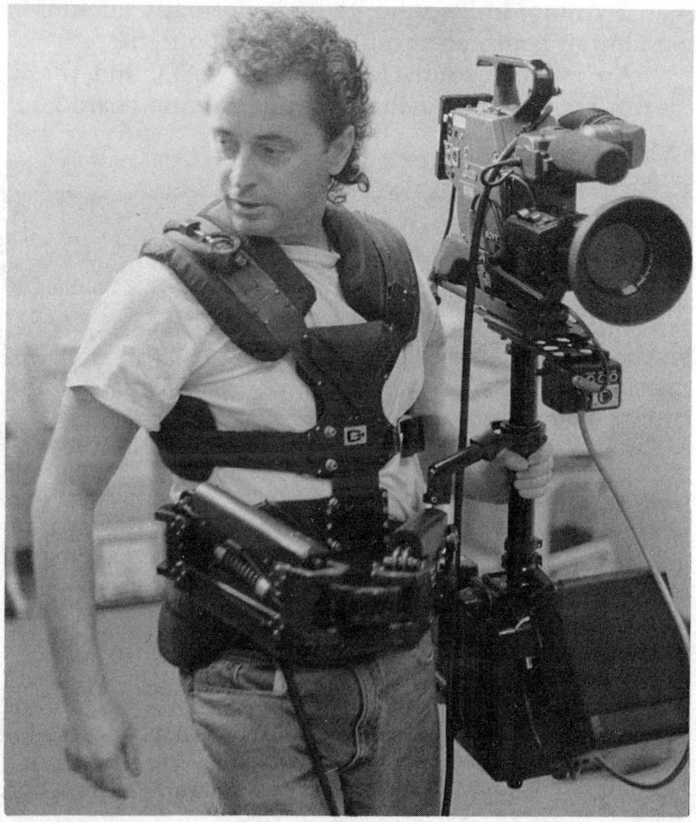

Figure 4.33 Steadicam unit in use on a corporate shoot. Units such as this allow more camera movement than a dolly and still provide very smooth motion.

Studio production Studio production often involves multiple cameras. In this case, a set has been built and "dressed" in the days before production, and all lighting and sound setups have also been completed in advance.

As the shoot gets under way, the actors are brought to the set from their dressing rooms. They rehearse with the director (Figure 4.34) and walk through blocking. At the same time, camera operators rehearse their focal lengths, moves, and positions based on the director's previous instructions or a shot list he has provided.

When the director is satisfied that the actors' performances are polished enough to record and the camera operators are fully rehearsed, he typically retires to a master control room. This "booth," as it's often called, provides him with images displayed on a wall of monitors, from all the cameras on the floor.

As he now gives the action cue and the scene plays out, the director instructs the camera operator to line up the proper shots. At the same time he calls out to a technical director (TD) to switch signals at the proper times. In this way, the selected images are recorded in sequence and the director can record all coverage of the scene—the master shot, medium shots, and close-ups—in a single good pass (Figure 4.35). When this has been accomplished, the next scene is rehearsed, and so on.

Audio production Audio production can take place either on location or in a studio. In a studio, there is normally an audio control room and soundproof booth arrangement for this purpose.

An audio recordist/engineer, the AD, and the director sit in the control booth. They utilize a sound mixing board and usually a ¼-inch

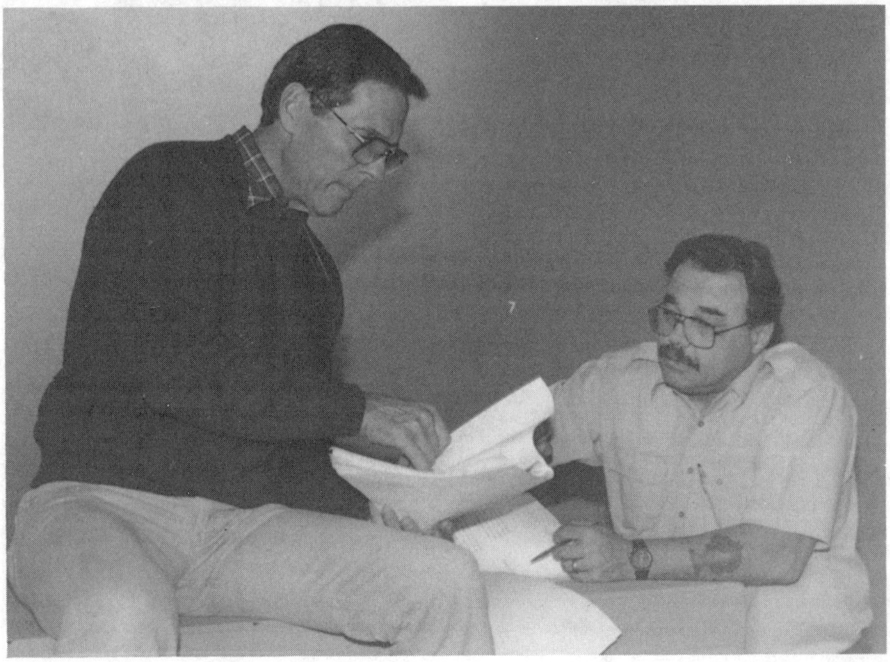

Figure 4.34 Director (right) rehearses with talent in the studio.

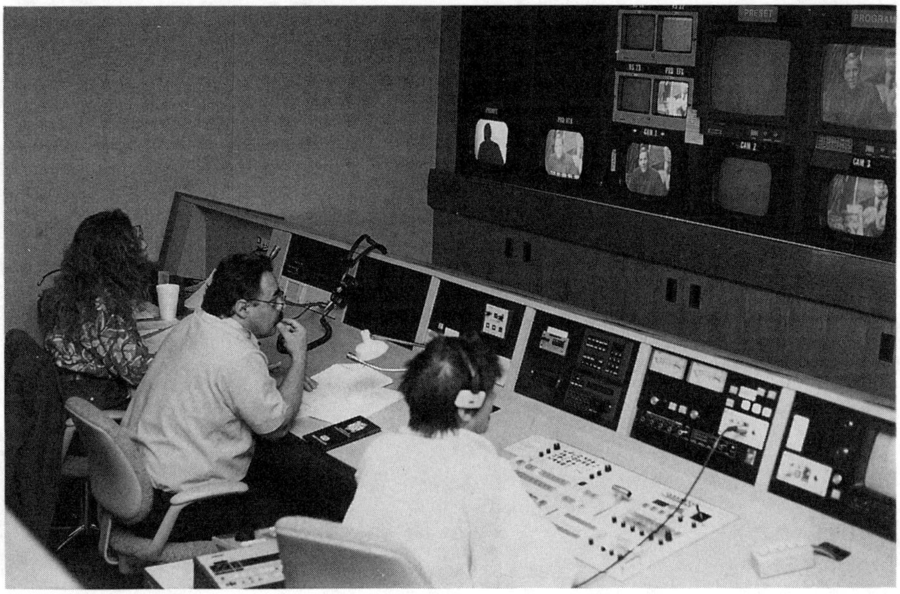

Figure 4.35 A live shoot in production. Seated in this control booth are, left to right, the AD, the director, and the TD. Monitors show various camera angles.

audiotape reel-to-reel recording system. The voice-over actor or "talent" sits in the soundproof booth and performs on the director's or engineer's hand signal, and multiple takes are rehearsed and recorded just as with the video portion of the program.

On location, audio recording is often accomplished in a quiet room or area. Instead of using ¼-inch audiotape, sound is recorded onto videotape by using the camera as the recording system but disregarding the visual element. For instance, the camera might be pointed at a slate for that scene and put into "record." It is left in this position as the actor reads his lines, and, later, in editing, only the audio portion of the recording is usable.

Postproduction

When all sound and picture elements of the script are recorded, production is "wrapped." The project then moves into postproduction or editing. Normally, corporate programming uses two editing processes: a rough cut or off-line and a fine cut or on-line.

The off-line Off-line editing is usually carried out on a comparatively simple editing system, such as the one in Figure 4.36. It has two purposes. The first is to create a rough cut of the program. This is accomplished by editing duplicate, work-print copies of the buy takes into a rough but accurate version of the show. A viewing of the rough cut is arranged, and the client and producer then approve or suggest changes if appropriate.

The second purpose of the off-line edit is to create an edit decisions list (EDL). The EDL is a list of time code numbers that reflect the exact edit points chosen by the off-line editor (Figure 4.37). Every segment of video or audio in the rough cut has a unique time code "address," and

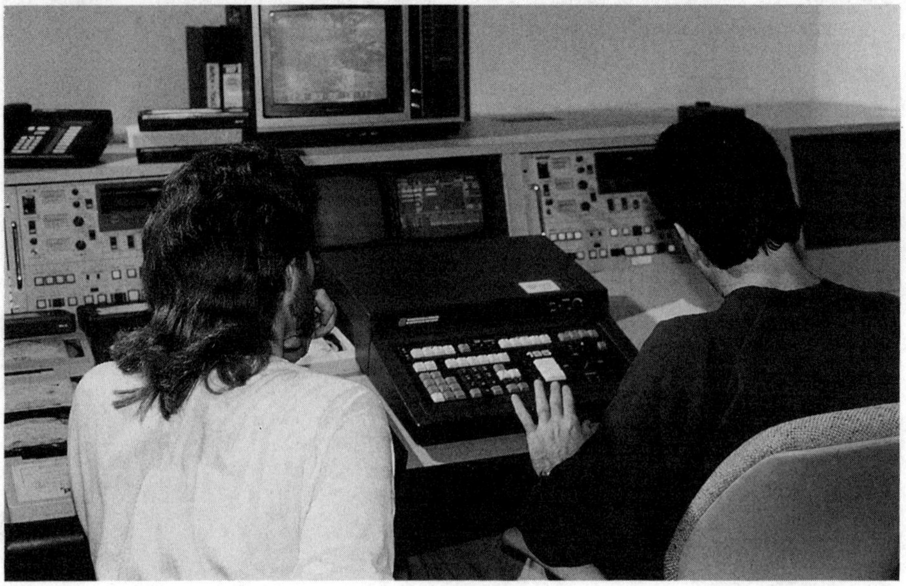

Figure 4.36 Off-line editing system in use. Controller is between editor (right) and director. VTRs are built into shelf.

these addresses correspond exactly to the identical segments on the high-quality original footage.

Once client and producer changes have been made, the EDL is revised and "cleaned" to reflect the exact "in" and "out" address points of every edit. This list of final addresses can then be used as the backbone of the on-line editing process.

The on-line The on-line is the final edit. It involves using the time code EDL created in the off-line to edit the original, high-quality segments of that same footage into what is called an "edited master."

The on-line is accomplished by loading the cleaned EDL into a much more sophisticated editing system, as seen in Figure 4.38. Typically, the EDL is inputted via floppy disk and used as a control tool to cue the operation of VTRs containing the original footage reels.

Because the on-line system cues VTRs according to the EDL and has the ability to control multiple VTRs as well as "outboard" or periphery equipment, much of an on-line edit can often be carried out very rapidly in an "auto assemble" mode.

A video switcher, such as the one seen in Figure 4.39, is also used in the on-line. The switcher is a powerful tool that adds visual effects like dissolves and wipes and provides many other functions. Digital video effects (DVE) "moves" are also included at this point with a special-effects box, such as the Abekas A53-D seen in Figure 4.40. Titles are also added, using a character generator (Figure 4.41), and music is typically included at this point as well.

Audio "sweetening" In some programs, more than just simple music insertions are required. Many times, the audio recordings must be en-

Reflection Problems Case Study 2

I once directed a scene in which the camera, mounted high on a crane, was to lower to eye level as a car driven by the lead actor and his wife was driven into the frame.

The idea was to see their car approaching from down the street, and then drop slowly to their level to gain a perspective on their conversation.

We recorded the shot on a sunny day, and I watched each take in a video monitor. Although the sun's reflection made the images on the monitors, a clear reflection of the entire setup became visible on the shiny black surface of the car's door.

Later, however, when we had returned to the studio and were watching dailies, I got a surprise. As the seat of the crane—containing the camera, the DP, and a focus puller—dropped to eye level with the actors, a clear reflection of the entire setup became visible on the shiny black surface of the car's door.

Although the group remained basically still and the viewer's attention was drawn to the actors, I realized that the reflection would probably make the shot unusable.

In the editing session, however, a switcher effect saved the day. I requested that the editor use the video switcher to create a black matte that matched the color of the car's door. As we edited the shot, when the car drove into the frame, the editor slowly faded the matte in over the reflection. Because it happened slowly and the color matched almost perfectly, no one was ever the wiser.

---

hanced or changed. Sound effects may also be required as may tonal adjustments to certain audio recordings.

This would be the case, for instance, when one part of a scene includes voice-over dialogue that is supposed to be the distant end of a telephone conversation. To make this conversation sound real, the "distant" voice, which was recorded in the field like all other sound segments, is equalized or "EQ'd" to give it the distinct, hollow tone of someone speaking on a telephone.

When such sweetening is required, the audio tracks from the edited master of the program created in the on-line must be "laid down" onto a special multitrack audiotape, using an audio recorder such as the one in Figure 4.42. All recorded sounds are then enhanced using multiple channels and an assortment of sound manipulation gear, including a multitrack board, such as that shown in Figure 4.43. Finally, the various enhanced recordings are "mixed down" typically to two tracks—one with narration and dialogue and the other with music and effects.

Once this sweetening is accomplished, the now-mixed-down audio tracks are "laid back" onto the original edited master of the program in sync with the pictures. At this point, the show is nearly complete.

CMX Ootimizer Wed Mar 04 11:43:16 1992 Page 1
FILE: 4061H.TXT

TITLE: RIDE SHARE
FCM: DROP FRAME

Edit	Reel#	Type	Tran	Source In	Source Out	Record In	Record Out
001	6	B	C	06:14:36:28	06:14:55:18	01:00:00:00	01:00:18:20
002	6	B	C	06:19:05:22	06:19:17:05	01:00:18:20	01:00:30:03
003	6	A	C	06:20:54:20	06:20:56:22	01:00:30:03	01:00:32:05
004	6	V	C	06:20:54:20	06:20:55:21	01:00:30:03	01:00:31:04
005	6	V	C	06:22:12:03	06:22:15:23	01:00:31:04	01:00:34:24
006	5	B	C	05:29:10:20	05:29:16:27	01:00:34:24	01:00:41:01
007	5	B	C	05:04:06:04	05:04:19:10	01:00:41:01	01:00:54:07
008	5	A	C	05:19:34:07	05:19:43:07	01:00:54:07	01:01:03:09
009	5	V	C	05:19:34:07	05:19:48:07	01:00:54:07	01:01:08:09
010	2	A2	C	02:20:15:02	02:20:16:05	01:00:57:10	01:00:58:13
011	5	A	C	05:22:18:18	05:22:21:07	01:01:17:23	01:01:20:12
012	5	V	C	05:22:18:18	05:22:27:27	01:01:17:23	01:01:27:02
013	2	A	C	02:05:42:11	02:05:44:10	01:01:20:12	01:01:22:11
014	6	A2	C	06:09:59:20	06:10:07:03	01:01:27:00	01:01:34:13
015	6	V	C	06:09:27:25	06:09:41:01	01:01:27:02	01:01:40:08
016	2	A	C	02:06:18:27	02:06:23:14	01:01:29:15	01:01:34:02
017	6	V	C	06:09:41:02	06:09:46:06	01:01:40:08	01:01:45:12
018	4	B	C	04:12:44:20	04:12:55:12	01:01:52:22	01:02:03:16
019	4	A2	C	04:19:43:26	04:19:52:20	01:01:56:06	01:02:05:02
020	4	A	C	04:14:38:24	04:14:42:01	01:02:03:16	01:02:06:23
021	4	V	C	04:14:38:24	04:14:40:18	01:02:03:16	01:02:05:10
022	4	V	C	04:19:04:15	04:19:10:19	01:02:05:10	01:02:11:14
023	4	A	C	04:19:05:28	04:19:10:19	01:02:06:23	01:02:11:14
024	4	A2	C	04:19:10:26	04:19:16:11	01:02:11:10	01:02:16:25
025	4	B	C	04:14:13:06	04:14:18:17	01:02:11:14	01:02:16:25
026	4	B	C	04:21:30:11	04:21:38:21	01:02:16:25	01:02:25:05
027	4	A2	C	04:21:38:20	04:21:44:22	01:02:24:17	01:02:30:19
028	4	A	C	04:17:50:20	04:17:52:21	01:02:25:05	01:02:27:06
029	4	V	C	04:17:50:20	04:17:57:21	01:02:25:05	01:02:32:06
030	5	A	C	05:14:04:25	05:14:18:10	01:02:43:07	01:02:56:22
031	5	V	C	05:14:04:25	05:14:37:03	01:02:43:07	01:03:15:17
032	2	A2	C	02:14:01:02	02:14:04:27	01:02:51:22	01:02:55:17
033	2	A2	C	02:14:06:00	02:14:24:15	01:02:55:17	01:03:14:04
034	AX	A	C	00:00:00:00	00:00:16:08	01:02:56:22	01:03:13:02
035	5	V	C	05:14:37:04	05:14:42:04	01:03:15:17	01:03:20:17
036	1	B	C	01:05:34:18	01:05:55:15	01:03:24:28	01:03:45:25
037	1	A	C	00:04:47:27	00:05:09:07	01:03:45:25	01:04:07:05
038	1	V	C	00:04:47:27	00:05:10:08	01:03:45:25	01:04:08:06
039	7	B	C	07:06:39:04	07:06:43:20	01:04:09:05	01:04:13:21
040	7	A2	C	07:03:11:29	07:04:26:00	01:04:13:19	01:05:27:20
041	8	B	C	08:19:48:05	08:19:57:16	01:04:13:21	01:04:23:02
042	9	B	C	09:14:13:29	09:14:22:06	01:04:23:02	01:04:31:09
043	3	B	C	03:01:31:00	03:01:34:15	01:04:31:09	01:04:34:24
044	3	B	C	03:03:05:00	03:03:08:15	01:04:34:24	01:04:38:09
045	3	B	C	03:12:29:17	03:12:33:02	01:04:38:09	01:04:41:24
046	8	B	C	08:20:07:10	08:20:18:11	01:04:41:24	01:04:52:25
047	9	B	C	09:16:14:08	09:16:25:18	01:04:52:25	01:05:04:07
048	7	B	C	07:04:00:17	07:04:03:17	01:05:04:07	01:05:07:07
049	7	B	C	07:06:40:22	07:06:43:23	01:05:07:07	01:05:10:08
050	8	B	C	08:20:41:11	08:20:45:25	01:05:10:08	01:05:14:22
051	3	B	C	03:04:56:27	03:04:59:13	01:05:14:22	01:05:17:08
052	9	B	C	09:13:45:24	09:13:51:13	01:05:17:08	01:05:22:27

A

Figure 4.37 Edit decision list (EDL) shows time-coded reference points of all edits or "decisions."

As a final step, a duplication master or "dupe master" is created on which the two previous audio tracks are mixed a final time or "summed" to one channel. This dupe master is then used to make the distribution copies that are handed to the client for her showings.

Production People

As you've now read, the television production process is not a simple one. On the contrary, it is a highly complicated and exacting profession that requires skilled creative and technical craftspeople.

CMX Optimizer Wed Mar 04 11:43:16 1992 Page 2
FILE: 4061H.TXT

Edit	Reel#	Type	Tran	Source In	Source Out	Record In	Record Out
053	9	B	C	09:13:52:11	09:13:57:11	01:05:22:27	01:05:27:27
054	7	B	C	07:12:26:04	07:12:32:21	01:05:30:24	01:05:37:11
055	7	AA/V	C	07:04:40:00	07:04:51:23	01:05:37:11	01:05:49:04
056	8	A	C	08:17:52:28	08:18:03:10	01:05:49:04	01:05:59:14
057	9	V	C	09:23:06:16	09:23:07:27	01:05:49:04	01:05:50:15
058	2	A2	C	02:19:23:25	02:19:34:01	01:05:49:08	01:05:59:14
059	8	V	C	08:17:54:09	08:18:03:10	01:05:50:15	01:05:59:14
060	7	A2	C	07:04:40:00	07:04:49:26	01:05:59:14	01:06:09:12
061	9	B	C	09:03:48:23	09:03:58:19	01:05:59:14	01:06:09:12
062	4	B	C	04:02:10:08	04:02:26:06	01:06:09:12	01:06:25:10
063	3	B	C	03:12:29:29	03:12:34:29	01:06:27:10	01:06:32:10
064	3	A2	C	03:09:46:05	03:10:09:02	01:06:32:09	01:06:55:06
065	8	B	C	08:13:15:29	08:13:38:24	01:06:32:10	01:06:55:05
066	3	B	C	03:03:23:00	03:03:28:13	01:06:55:05	01:07:00:20
067	3	A2	C	03:09:55:00	03:10:09:11	01:07:00:19	01:07:15:00
068	9	A	C	09:20:02:28	09:20:12:22	01:07:00:20	01:07:10:14
069	9	V	C	09:20:02:28	09:20:17:22	01:07:00:20	01:07:15:14
070	3	A2/V	C	03:11:00:02	03:11:11:14	01:07:19:08	01:07:30:20
071	2	A	C	02:04:10:02	02:04:11:07	01:07:21:08	01:07:22:13
072	2	A	C	02:23:27:28	02:23:36:05	01:07:22:13	01:07:30:20
073	7	A2	C	07:12:24:12	07:12:42:29	01:07:30:20	01:07:49:07
074	8	A	C	08:30:49:21	08:31:03:10	01:07:30:20	01:07:44:07
075	8	V	C	08:30:49:21	08:31:08:10	01:07:30:20	01:07:49:07

B

The following pages contain a listing of the people who are involved in this process and a brief description of the role each plays. Like the equipment and the process itself, the director must have a working knowledge of each of these job functions and how they affect his own endeavors.

The Producer

The producer oversees the entire production process, from the day the client makes the request until the final program is completed. The producer supervises all production activities and guides the production according to her own business and aesthetic preferences.

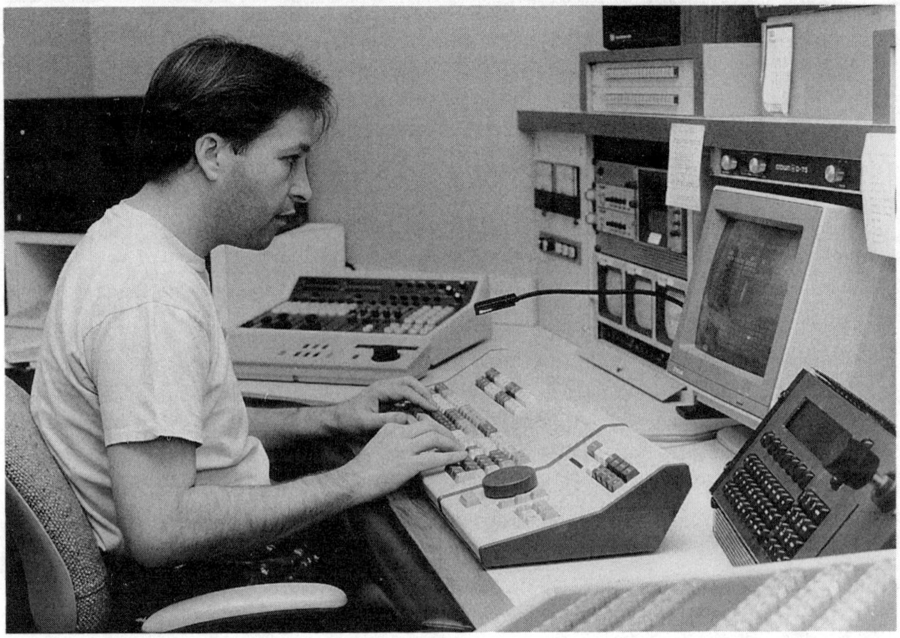

Figure 4.38 On-line editing system in use.

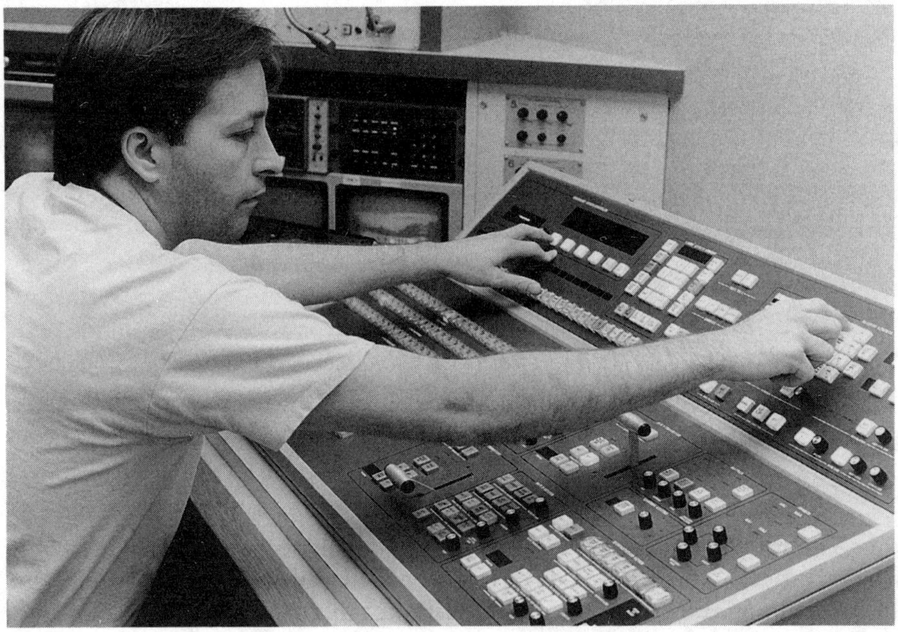

Figure 4.39 Video switcher manipulates video signals in numerous ways. This Grass Valley 200 provides wipes, fades, dissolves, and other effects. It also generates backgrounds, combines pictures, superimposes titles, and performs many other functions.

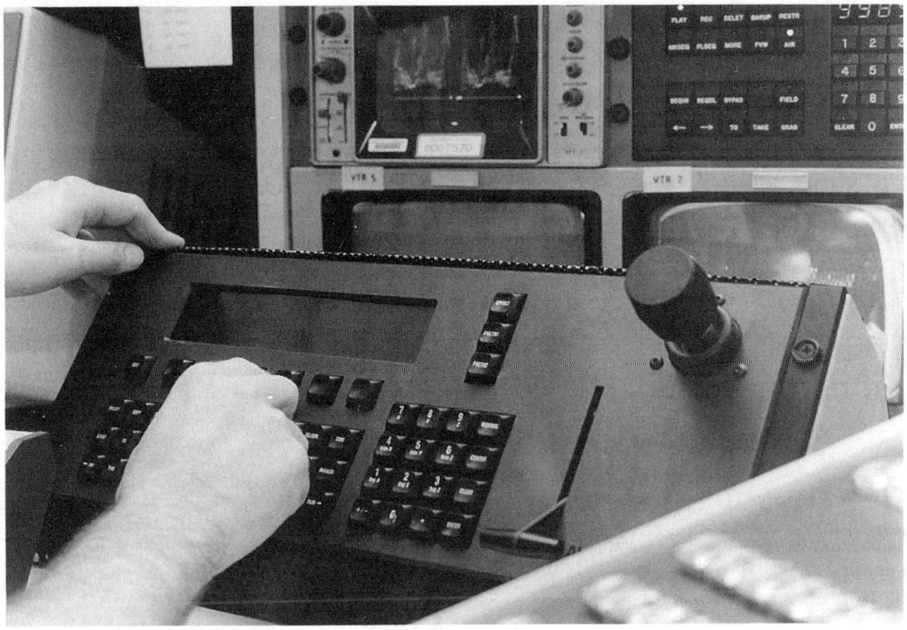

Figure 4.40 An Abekas A53-D digital video effects generator or "DVE box." Such units provide even more control and manipulation power than video switchers.

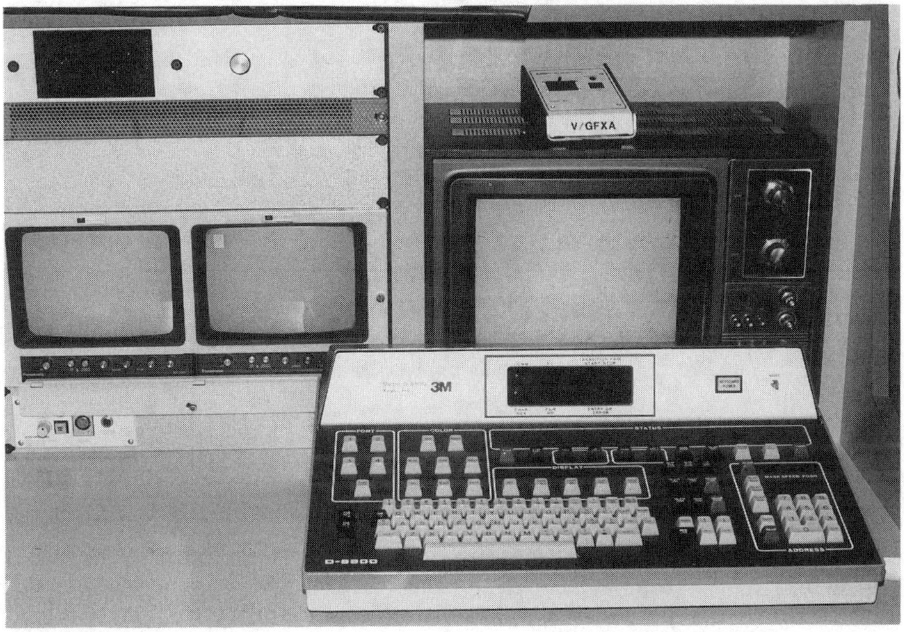

Figure 4.41 Character generator creates titles to appear on screen. Numerous styles and fonts are available. The unit also provides simple graphic animation.

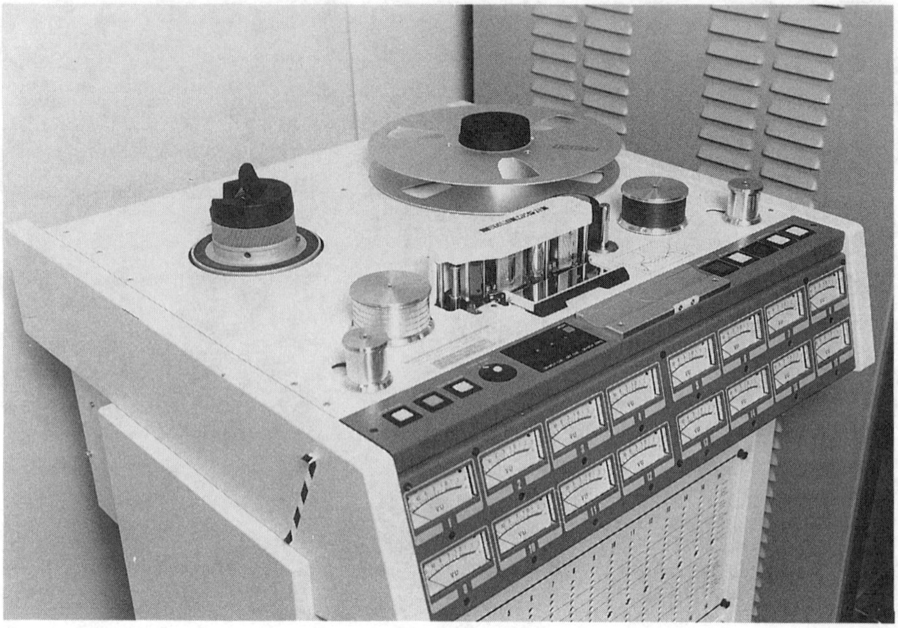

Figure 4.42 Multichannel audio recorder. This unit will record up to 16 individual tracks. These can then be mixed down to a single track.

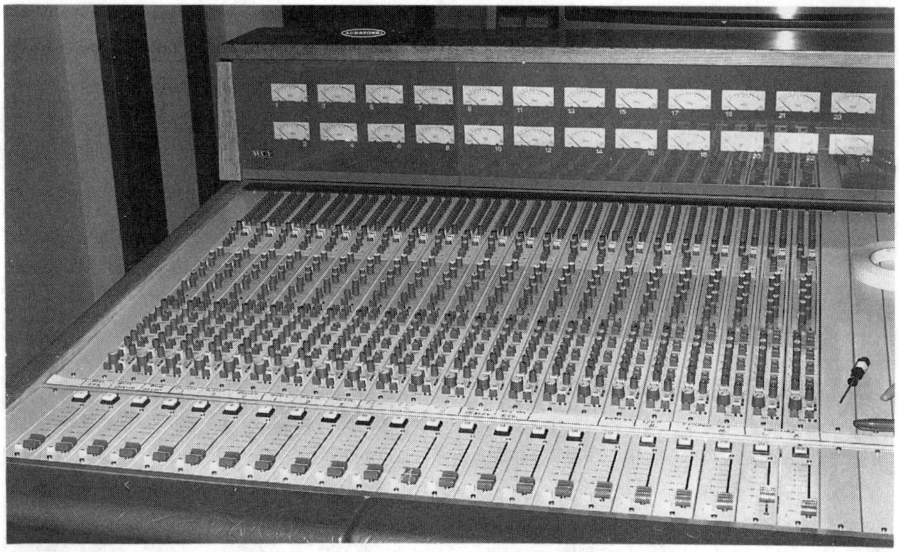

Figure 4.43 Multitrack audio board provides control and manipulation of individual audio signals being fed to audio recorder.

The Client

The client is the person who places the responsibility for her communication need in the producer's hands. The client also frequently pays for the production and has input into technical as well as aesthetic aspects of her program.

The Director

The director is responsible to the producer and the client. Her job is to record all video and audio elements in a way that makes them appropriate for editing into a completed program. The director supervises the cast and crew and bears the ultimate responsibility to the producer for all recorded segments. The director has the greatest aesthetic control over the production, since she works with the actors, director of photography, and sound recordist to create recorded segments according to her detailed visualization.

> *As a director, you needn't be an expert at lighting, sound, stage design, and so on. You should have a clear vision of how your scene should play on the screen, and then utilize the talents of your production team to create that vision.*
>
> —Ralph Phillips
> Writer/Director

The Assistant Director (AD)

The AD is the director's right-hand person. He helps arrange all aspects of preproduction and coordinates client, crew, and talent on location or in the studio. Most times in corporate productions, the AD also takes script notes and video shot logs. In preproduction, the AD relieves the director of many tasks, allowing her to focus on script interpretation, talent selection and rehearsal, location selection, scheduling, and handling of client and producer.

The Production Assistant (PA)

The PA works primarily for the AD or the director. The PA handles telephone calls, equipment rentals, prop deliveries, mail, facsimiles, personal pickups and deliveries, and many other important production details. The PA may also work for the producer at times, doing filing or runner-type activities.

The Director of Photography (DP) or Videographer

The DP is responsible to the director for the entire visual look of the program, including all camera work and lighting. The DP is the director's closest aesthetic ally, since he understands what the director is attempting to achieve visually and uses his many photographic skills to help achieve this end.

The Camera Operator

The camera operator operates a camera but is not responsible for lighting. On corporate location shoots, the DP usually also operates the single camera. On multicamera studio shoots, the DP operates one camera and camera operators operate the others.

The Gaffer

The gaffer is the DP's main assistant. She sets up lights, checks amperage loads, verifies fuses and breakers, connects generator cables, and basically provides constant lighting and electric help. On corporate shoots, gaffers also frequently act as camera assistants, "pulling focus" when necessary, changing batteries, setting up tripods, and attaching other accessories.

The Grip

The grip is a general labor assistant. He does most of his work for the gaffer. He carries cables and sets up "C" stands, flags, silks, and nets. The grip also does a great deal of loading and unloading of tools and equipment during setups and strikes. A dolly grip pushes or pulls the dolly on which the DP and sometimes an assistant ride.

The Sound Recordist/Mixer

This is the person responsible to the director for all sound recordings made on the program. The sound recordist—or audio person, as she is normally called—suggests the proper mics and placement and ensures that the recorded pieces are of the best quality and consistency under the circumstances. She also operates a mixer to ensure that sound levels are consistent in terms of tone and volume.

The Audio Boom Operator

The audio boom operator is the sound recordist's assistant. He attaches the mics or holds them on a boom in position during recordings. He also maintains mics and batteries to ensure that they are in good working order when needed.

The Floor Manager or Stage Manager

On studio shoots, when several cameras and a number of actors and crew members are working in unison, a coordinator is needed to act as the director's voice on the studio floor. The floor manager carries out this responsibility. At the director's request, she cues talent, gives start and stop signals, and handles any general questions or problems that may arise during production.

Teleprompter Operator

A teleprompter is a word processing unit that projects a host's narration onto a translucent pane of glass mounted over the camera's lens. The host can thus appear to be speaking from memory as he looks directly into the camera and reads his lines. The teleprompter operator usually inputs the script text and controls its scrolling action in production, thus ensuring that the words the host is reading move along at the proper pace.

The Makeup Artist

Makeup artists are not a part of most corporate shoots. They are hired from time to time, however, especially when the production involves a number of actors or celebrities. In corporate programs, makeup is often handled by the actors themselves and minor touch-ups are taken care of by the floor manager, AD, or PA.

The Off-line Editor

The off-line editor works closely with the director immediately following production. She uses work prints of the footage to create a "rough cut" of the program. The director may sit in on these edit sessions to verify that the scenes are playing out as he had envisioned. In some corporate situations the director and producer may sit with the off-line editor at the outset of the edit session, give her specific instructions, and then allow her to independently edit portions of the show. Periodically, her work is reviewed and changes are made as needed.

The On-line Editor

The on-line editor performs the final edit. When he begins work on a project, most of the aesthetic decisions relating to composition, pace, and visual continuity have been made. The on-line editor, however, adds the final touches, which often enhance the show greatly. These include dissolves, wipes, fades, sound effects, music, titles, and DVE.

The Audio Sweetening Engineer

The audio sweetening engineer or postaudio engineer is highly skilled at incorporating and manipulating multiple sound elements into an existing sound track. She is also skilled at mixing down her work onto only one or two audio channels. She adds music, sound effects, and various enhancements to the existing recorded sound to greatly enrich the program's final sound track.

Finally . . .

At this point, you should have a clear understanding of how corporate television programs are made, the people and equipment involved in that process, and many of the dos and don'ts of working in the corporate world. As mentioned earlier, this combined knowledge will act as a solid foundation that comes into play on every directing project.

The skills we will now begin to explore are the *framework* that stands on that foundation. Like the foundation itself, these skills too are integral to every directing assignment. Unlike the foundation, however, these framework skills are not passive. They manifest themselves in the form of *actions* a director must take continually, during the entire span of his project.

5 Script Aesthetics

In Chapter 1 we established a definition of what a director *is*. In Chapters 2, 3, and 4 we looked closely at her production tools and the basic processes and environments within which she operates. With this knowledge in place as a foundation, we are now equipped to take the next logical step and examine what exactly a director *does*.

Illusionism: The Director's Art

Webster's Collegiate Dictionary defines the word *illusionism* as follows:

> the use of artistic techniques (as perspective or shading) to create illusions of reality

This phrase describes very closely what a corporate director does. To make the definition fully correct, we would revise it to read:

> the use of artistic techniques (as performance, videography, and others) to create illusions of reality in a media presentation

The illusions referred to here are things like two characters cast as sales reps discussing a new technique over lunch; or a host cast as a first-line supervisor presenting customer-handling tips directly to us as viewers; or an employee working at his craft as if there has been no camera, lighting, or directorial setup of any kind.

When directing a CEO, be sure the executive is given your full attention. Also, you must exhibit a nonabrasive self-confidence in your abilities as a director.

—Dan F. Johnson, M.A.
Producer/Director

In order to make these media illusions appear to be unrehearsed moments captured on the screen and at the same time to engage a viewer's interest, the director works with five aesthetic elements:

script aesthetics
human aesthetics
audiovisual aesthetics
technical aesthetics
general aesthetics

Each of these elements has a profound effect on the illusory credibility of the director's work. In fact, even the most subtle aesthetic faults can destroy the credibility of a shot, scene, or entire program on the spot. Keep in mind as we discuss them, however, that the critical importance of each one can sometimes be deceptively unassuming. Indeed, many manifestations of the aesthetic qualities we are about to discuss can be extremely easy to misjudge or overlook. As we will see, the shaping and

Picture Perfect Case Study 3

I once knew a director who had a somewhat unbalanced view of the different aesthetic qualities of his work.

He had been a camera operator and a director of photography before he began directing. As a result of this background, he was often so concerned with the visual aesthetics of the programs he made that he sometimes let this override his judgment on human aesthetic problems that should have been corrected.

If a dolly or some other camera move he was attempting to achieve happened just right but the acting was so-so, he would accept the scene based on its audiovisual quality alone.

As a result, his shows tended to have very polished videographic segments, consisting of nicely framed images and lots of attractive dolly moves and lighting. The performances of his actors, however, frequently left something to be desired.

The problem with this, as he eventually came to realize, is that beautiful pictures will not convince an audience to accept less than credible performances. In the end, the acting *must* be top-notch or the program will suffer—as this director found out in several embarrassing rough-cut meetings.

Eventually, he realized this fault and began placing increased emphasis on casting and achieving solid performances. As he gained a more balanced perspective of his work, his programs immediately began receiving better acceptance.

---

guiding of these aesthetic subtleties is one facet that makes the director's work so challenging and at times so risky.

Script Aesthetics

A script is a transitional document. A writer creates the script as a detailed representation of how his vision should appear on the screen. Since the writer has worked with a client and producer while creating the script and since his work has probably been approved by both these individuals, we can assume that they too share and fully agree with the writer's vision.

If the director is to successfully transform the script into live action on the screen, her first task is to carefully *analyze* it and gain an intimate and multifaceted understanding of what the writer has attempted to communicate. An intimate understanding of these elements is critical on the director's part because she could never hope to guide actors, videography, preparations, editing sessions, and so on without making specific judgments as to how the material should play in front of the camera. Those specific judgments cannot be made without an intimate understanding and analysis of every nuance in the script.

In feature film and prime-time television production, that analysis means careful exploration of characters and their development, plot structure, conflicts, crisis tension levels, and many other dramatic elements. In corporate television, complicated plots, pronounced character development, and intense emotional conflicts are not often used. When they are used, they are often much less complex than those in dramatic presentations. This is because, as we have learned, corporate television programs are not made strictly to entertain or gain a viewer's dramatic involvement. They may contain some degree of these elements, but they are produced primarily to convey either specific information, such as five steps on how to make a sale, or a general concept that will motivate viewers to follow those five steps.

This does not change the fact that a corporate director must analyze the writer's script; it simply means that her analysis is usually focused on somewhat different priorities. These include:

informational clarity
character profiles
plot and structure
tone
pace

Informational Clarity

On any directing project, one of the first questions a director must ask herself is this: If shot as the writer has described, will this script create for its *viewers* the same informational clarity it has for its *readers*—the client, producer, and any technical experts who may have read and approved it? In other words, will it work on the *screen* that same way it works on the *page*? She makes this assessment by reading the script over carefully scene by scene, applying her experience and judgment as a director of visual media and continually asking that same question.

If the answer continues to come up "yes" with each scene, she can go about the rest of her script analysis feeling comfortable that from an informational standpoint the script is aesthetically sound. Most times, this will be the case, since, as we have said, the producer and client have no doubt carefully scrutinized and revised the script in terms of content. If, at certain places, however, the answer comes up "no," she must bring those areas that fall short, together with her suggested solutions to those shortcomings, to the attention of the producer and client.

For instance, let's assume that a director is reading a script that she feels is informationally sound and she suddenly comes across one section that appears problematic. The information being communicated in that section pertains to two specific aspects of quality service: *focusing* on the customer's problem and being *courteous.*

The troubling section looks like this:

```
INT AIRPORT TERMINAL—DAY
John walks up to the ticket counter. The TICKET PERSON
approaches with a courteous smile.
```

NARRATOR (VO)
Quality service is important in any job because it's critical to customer satisfaction.

The ticket person takes John's ticket, *focuses* on his problem, remains *courteous*, keys the information into his computer, makes a change to the ticket, and gives it back to John, corrected. John exits the shot perfectly happy. He has just had an encounter that displayed these two specifics of quality service.

DISSOLVE

At first glance, this scene might appear to be effective. The ticket person seems to act in a courteous and focused manner, and the scene certainly talks about focus and courtesy. The writer even closes the scene

Script Interpretation Case Study 4

I was once given a corporate script to direct that I felt lacked basic structure and informational clarity. After reading it over several times and visualizing how it would play out on the screen, I was convinced that, if shot as written, it would not do the job.

I was faced with the dilemma of how to express these feelings to the producer and ask for what I was sure would have to be an extensive rewrite.

My first concern was to *not* come off like a prima donna or give the impression that, because she had approved the script, her judgment was poor. I was equally concerned, however, with the thought of directing a program that I felt sure would fail.

After wrestling with this problem for several hours, I went to the producer and prefaced my comments generally like this:

> I've read this script over, and I have to be frank with you: I feel it has some significant problems. You and I have worked together before, and I respect you. I also feel I can talk to you frankly about this, because I'm sure you realize that I'm concerned with the good of the project and with ensuring that, if you spend the money to do this show, it will be effective. I'd like to talk over the problems with you and suggest a rewrite. If, after you've heard my reasoning, you'd rather shoot the program as it's scripted, I'll certainly do it for you and give it my best.

Exactly what I had hoped would happen *did* happen. She heard me out, was grateful for my concern and positive criticism, and obtained a complete rewrite of the project before I directed it.

In the end, the program was a great success, and everyone, including me and the producer, was happy.

by confirming this with the line "He has just had an encounter that displayed these two specifics of quality service."

But *viewers* will never read the writer's scene description, including his closing line. They will only *see* the ticket person approach wearing a smile and go through the motions of writing out John's ticket. What they will *hear* will be the scripted narration.

Given these two elements, will they understand that focus and courtesy are the key informational elements being displayed on the screen? Probably not. They will undoubtedly get a general sense of quality handling from this scene, but no knowledge of the specifics. This is because although the writer has included the key items in his scene description, he has not included action or dialogue to support the message.

What, then, is the director's solution to this type of problem? In this case, there are several possibilities.

One solution would be to add action and dialogue that would display a sense of focus and courtesy. For instance, several telephone calls or pesky fellow workers might act as distractions to the ticket person. If it became *visually clear* that she was consciously ignoring these distractions to keep her *focus* on her customer, the point would be made.

The courtesy specific could also be conveyed through additional dialogue and action, such as several courteous lines delivered by the ticket person. Another possibility would be to add narration and perhaps superimposed titles. For instance, the narration might be rewritten this way:

```
                NARRATOR (VO)
Quality service is important in any job
because it's critical to customer satisfac-
tion. In this case, even in the face of
constant distractions, the ticket rep
keeps her attention focused squarely on
John's problem. She also remains cour-
teous every step of the way.
```

This additional narration would certainly help make the point. If the scene were also supplemented with *superimposed titles* that said "Focus" and "Courtesy" at the proper places in the narration, its informational clarity would be further enhanced.

A rule of thumb This is only one of many types of informational problem areas a director may have to work out, but in each case the general rules remain the same:

- Analyze each scene.
- Determine if it is informationally sound, not from a print standpoint but from an *audiovisual* one.
- If not, determine what revisions are needed.
- Revise or suggest revisions to include *action, dialogue, narration,* and, where appropriate, support *titles.*

These suggested revisions should be presented to the producer and possibly the client at the outset of the director's work. If the producer agrees with the suggested changes, they can be made, often by the original writer, and the project can move forward. If the producer disagrees, the director must either disregard them herself or decide whether she wants to continue with the project—assuming that it may not meet its objectives in the end.

Usually, if the problems are fairly minor, a good course of action is to go with the producer's wishes but perhaps find some way to work with the existing scenes to enhance them. If the director feels the problems are major enough to have a terminal effect on the entire project, she may want to remove herself by having a frank but tactful discussion to that effect with the producer.

> *The director, like the writer, must care about the audience.*
>
> —Patti Ryan
> Producer/Director

Character Profiles

Characters are the human elements in a program and thus the most important to audience identification. The director's challenge is to make her characters appear real and engaging. In order to do this, she must get to know them intimately through her script analysis. In our chapters on human aesthetics and interpersonal skills, we will explore these human aspects of the director's work in much greater detail.

Strictly in terms of the script, however, the director should think of character profiles in terms of two important elements: the *appearances* and *personalities* of the actors who will appear in her program. If these two elements will help achieve the script's objectives, the director's chances of a successful program are greatly enhanced.

In corporate programs, the director will work with host and dramatic or humorous character profiles.

Host roles A host is a common role in corporate television. A host is a tour guide of sorts, leading the viewer through parts or all of the information to be communicated. The director should remember that although the host looks directly at the camera and talks to the audience, he is still a *character* and must be cast and played accordingly.

Most times, a host's *personality* should be friendly, warm, and conversational. His *appearance* should complement the content being delivered and the setting established in the script. For instance, if the content of the program concerns instruction on how to supervise service-order writers, the host role could be a supervisor of these types of employees. This would allow audience members to identify with the host character and view him as a credible source of information. For example:

FADE IN:

EXT TRUCK YARD—DAY

ESTABLISHING SHOT—A WIDE, HIGH ANGLE of the yard. A few trucks are pulling out for the day's routes. Others are backing into the dock. A group of trucks remains parked near the rear of the yard. CAMERA PUSHES IN to this area.

DISSOLVE

A CLOSER ANGLE—ON ONE TRUCK. Our HOST steps down from the cab of a shiny new semi. He is a "hard," middle-aged man dressed in jeans and a plaid shirt. He wears a San Francisco Forty-niners ball cap. He steps toward the CAMERA, gestures back over his shoulder at the truck, and says:

HOST
You and I both know that rig weighs close to a hundred thousand pounds. And we also know that stopping it on a dime isn't easy . . .

This writer's description of a fairly hard-looking man, probably in his thirties or forties, seems to create an appropriate appearance for this role. The character's personality would seem to be roadwise and no-nonsense but also considerate enough to want to pass on what he knows about truck safety to his viewers. He should probably appear to be in fairly good condition also, but there would be no need for him to be particularly handsome. In fact, if he were to appear tough but presentable, that would do fine. An overweight, underaged, or effeminate person cast in this role probably wouldn't work as well. Such as actor would seem out of character delivering the lines scripted in the surroundings called for.

Obviously, a host cast as, say, as bank teller or an executive would require different considerations on both the writer's and the director's part.

Dramatic or humorous roles Actors and actresses in role-play scenes have dramatic or humorous character profiles. Unlike a host, they act out their parts as if neither the camera nor the audience exists.

A good scriptwriter will have developed characters that work naturally to enhance the message being communicated. In doing so, he will have given them effective dialogue and actions that fit with their character profiles and at the same time move the script toward its objectives.

During her script analysis, the director must clearly visualize each of these characters in action on the screen and, as with a host character, cast accordingly. For example, if the script is humorous, perhaps she will want to cast actors with appearances or specific "looks" to enhance the laughs being delivered.

This next scene, from a program on customer handling for employees of an ice cream shop, illustrates one example:

INT ICE CREAM PARLOR—DAY
A typical summer morning in a J.J. Martinson's Ice Cream Shop. The clock says 8:30 A.M. The counter EMPLOYEE is preparing the day's sale tags. There is one very thin CUSTOMER standing near the counter licking a single scoop of vanilla.

HOST (VO)
At times, you may even have to deal with some, shall we say, odd customer requests . . .

Suddenly, a short, chubby man approaches the counter, smiles, and says:

CHUBBY MAN
I'll have four banana splits, two strawberry sundaes, and three half gallons of Cookies 'n' Cream, please.

As he says this, he hoists a large, picnic-type cooler up onto the counter. On its side, the word *Lunch* has been stenciled. Both the counter employee and the thin man stare, aghast at the request.

Obviously, a fat man would be appropriate for this role. He would also seem to have a jovial personality, either unconcerned with his weight problem or totally unaware of the outlandish nature of his request. As the director considers this character, he might want to cast the part to further embellish the role. For instance, he might want the character to appear unintelligent, perhaps in keeping with his actions. In this case, the director might visualize the character in Levi's, a T-shirt, and a baseball cap, perhaps on crooked.

On the other hand, what if the roles were reversed? Suppose the very *thin* man requested the cooler full of banana splits as the *fat* man stood by with a small, single cone? This "surprise" character profile might add a fresh and original quality to the humor that could make the scene not just funny but unforgettable. A character change of this significance, however, is something the director should suggest to the client and producer before actually casting the actors.

Then there is the counter employee. What should he or she look and act like—young? old? male? female? heavy? thin? nerdy?

As you can see, the choices can be numerous. The implications of those choices, as we have said, are *critical* to the director's end product; for if characters are miscast, the result is almost always distracting and less-than-credible performances that "stop the show" for the viewer and thus break the illusion.

For this reason, the director must go through the type of character profile analysis we have just explored with every character in her script. She must know what her characters should look like and how they would think and act in all situations established in the script. It is only with this knowledge that she will then be able to cast appropriately and, later, develop the right character traits and motivations during rehearsal and shooting.

Plot and Structure

Plot Few if any corporate programs have dramatic plots with the depth and complexity of feature films. This is the case for two reasons. First, corporate programs are usually too short to allow for adequate plot development. Second, as we've established, the focus of corporate programs is primarily on information or motivation rather than the things intrinsic to drama—character growth, story lines, conflicts, and resolutions.

Many times, however, fairly simple plots *are* used in corporate programming. I once directed a corporate program that used an employee and her spouse visiting a weekend company get-together to reveal information about the company and its people.

The plot was simple. As the couple arrived, it became obvious that the employee's husband was very uncomfortable. He was sure he would have absolutely nothing to talk about with all of his wife's "technical phone company" friends. His wife was irritated at his stubbornness and close-minded opinions of her job and fellow workers.

As the husband began to meet people, however, and find out that the company was really much "more than just a phone company," he found the conversation more and more engrossing. With each encounter, his negative attitude slowly diminished. By the end of the afternoon, he had learned that his spouse's company was involved in an assortment of fascinating business ventures and that the people he thought would be technical and stuffy were just like him—good-natured and friendly.

Although this may not have been high drama, it did involve specific plot conflicts, turning points, and character motivation and development. As the director, I had to pay careful attention to these elements during my script analysis in order to make the program ultimately work on the screen.

For instance, it was essential that the husband appear sluggish, depressed, and almost pouting in his opening scenes. This temperament had to develop into a mood of general acceptance by the program's midpoint, and in the final scenes he had to appear actually enthusiastic about his wife's employment and the possibility of socializing more with her friends.

The wife, on the other hand, had to begin in a cranky and upset mood. This had to change to a sense of hopeful anticipation as she watched her husband meet one friend after another. Finally, she had to end with a loving smile on her face as she realized her husband had come around.

Since the script was shot totally out of order, I had to be continually conscious of exactly where we were in the plot so that when the scenes were cut together later, this character development would work as scripted.

Thus, you can see that just as with character profiles, the director must know every plot element in her program in great detail. She acquires this detail by carefully reading the script many times over. She continually looks for subtle changes in character, and she pays close attention to the overall story flow. As she does this, she asks herself questions such as the following:

Where does the story actually start?
What is the midway point?
What traits or occurrences motivate each character?
Are there obstacles or crises each character must overcome?

If so, what are those obstacles or crises?
Do the characters change in any way as a result of plot elements?
When do these changes occur, and what plot elements cause them?
Does the script have a definite "ending," such as a climax and resolution?

With these questions answered, the director will have a specific and thus functional grasp on the plot of her program.

Structure and transitions Corporate programs are often carefully structured to impart information in logical, simple-to-follow segments. Part of a simple structure is the inclusion of appropriate transitions between those segments.

Many times, this structural aspect replaces dramatic or humorous plot elements in corporate programs. This would be the case, for instance, when a host and general cover footage are used to impart information. In such a case, a detailed understanding of the structure of the narration and cover material replaces the director's need for the same knowledge about story elements.

In some cases, both story and informational structure become important. This would be the case in a script in which a host introduces a story and then returns from time to time to provide transitions to subsequent parts of the story.

Just as with the other aspects of her script analysis, the director must become closely attuned to these structural and transitional elements. This allows her to use the appropriate visual and sound techniques to help clarify the material for the viewer. If the structure of a script is confusing or unclear to the director and she does not resolve her confusion, she will have great difficulty making the script play well on the screen. One subtle example follows in this script segment:

INT STUDIO—LIMBO SET
We find the Host leaning against her desk in black limbo. She says:

HOST

So you can see that safety isn't always determined by the attitude of the individual or her mood. In this case, Helen was a good employee, her attitude was positive, and she was in a jovial mood. But she still got hurt badly. The story of John Bell, on the other hand, shows us that training does indeed play a part in developing safe working habits.

Although this segment is scripted as a single monologue, there are two separate thoughts here. The first is what appears to be the wrap-up of a story focusing on an employee named Helen. The second thought is the introduction of the new story on John Bell. This juncture is a good place for the director to think about doing one of several things that could enhance the program.

For instance, she might want to simply shoot this scene having the host turn to a new, closer angle immediately after the words "she still got hurt badly." This simple turning motion, combined with the closer angle, would emphasize the change of thought to John Bell and provide a comfortable visual transition for the viewer.

A second choice might be to dissolve away from the host to a shot or two of John Bell on the job as the host introduces him in voice-over. Again, this would add finality to the previous thought and move the viewer comfortably into the new segment.

When the director has carefully analyzed the script, she will find many of these structural stops, starts, turns, and signposts. Becoming aware of them herself, she will then be able to direct her program in a way that helps make the content perfectly logical and clear to her viewers.

The visual elements that eventually become the manifestations of these structural elements—such as wipes, dissolves, and DVE—are subjects we will cover in the chapter on audiovisual aesthetics. For the moment, let's move on to the last two aesthetic elements in the director's script analysis.

Tone

The tone of a script might be thought of as the mood it evokes. A script may contain several tones, all adding up to a single unifying or overall mood.

For instance, consider a script about an executive who falls prey to substance abuse, loses everything, and eventually recovers. It might start with a suspenseful tone as the executive is shown beginning his drug abuse. Later scenes might have sad or depressing tones as he descends into the horrible reality of what drugs do to individuals. Later still, the tone may be one of hope as he undergoes treatment. Finally, he may experience joy as he succeeds and returns home to his family. The overall tone or mood of such a script could be referred to as uplifting or inspirational.

During script analysis, the director must carefully analyze tonal elements—both in individual scenes and in the work overall. In doing so, she will be prepared later to make decisions that will help create these qualities as each scene is prepared and shot.

For instance, in the first part of the script just described, the director may want to heighten the suspenseful tone by using a hand-held camera that follows the executive into a dealer's house. Other oblique perspectives, such as low or overhead angles, would also increase this sense of tension.

During the sad or depressing parts, the director might request dark lighting with hard shadows. This would heighten the feeling of loneliness and depression. These scenes might also be shot suggesting nighttime.

Finally, in the closing inspirational segments the director might want bright, sunny scenes, perhaps long focal lengths in outside locations or even slow motion to hold and savor inspirational moments for the viewer.

A host program may not be dramatic, but, as we've established, a host is still a character and she too should project a sense of tone in her delivery. It might be positive, urgent, friendly, or abrupt. It may also be sad or inspiring. Just as with story and structure, host segments may also be used as transitions between story segments. In these cases, the director must simply consider the host segments like any other parts in the overall program.

In each of these cases, casting, props, coverage, music selections, scheduling, and a variety of other elements would be affected by these tonal script interpretations and decisions on the director's part.

Pace

Pace is linked to tone in that one often influences the other. Pace too can vary in different parts of the script. As with the other elements we've been discussing, the director must gain an accurate sense of the script's pace or varying paces. This sense will affect how she prepares and shoots the program, as well as how it will be edited.

Remembering that the director is constantly creating illusions, it might help to think of pace in terms of "life speeds." "Typical" could be considered a normal sense of how things happen around us all the time. A "faster" pace gives the viewer an illusion that events are happening more rapidly, and a "slower" pace tends to do the opposite.

With these ideas in mind, suppose the director is analyzing a script that teaches telephone installers one reason *not* to go into unattended backyards: Dogs may be present. She may find three scenes with totally different pacing:

EXT COFFEE SHOP—DAY
The installer leaves the coffee shop with a toothpick in her mouth and a smile on her face. She has just had lunch. She gets into her truck, picks up her order folder, and checks out her next stop. She puts the folder aside and slowly pulls out onto Fifth Street.

EXT HOUSE—DAY
Arriving at the house, the installer exits her truck, straps on her tools, and walks to the front door. She RINGS THE BELL several times but gets no answer. She is about to leave when she decides to peek over the fence to see if she can spy the trouble.

She moves to the fence, looks over it, and, sure enough, a phone wire has been torn free from beneath the eaves of the house. The wire is within 10 feet of her. She checks the gate latch. It's open. She peeks in and CALLS OUT.

INSTALLER
Hello! Anyone back here? Hello!

No answer. She hesitates, and tries again.

INSTALLER (CONT.)

Hello!

Still no answer. She decides to go in. As she closes the gate behind her and steps forward, out of nowhere a huge German shepherd charges around the corner. The installer dashes for the gate and reaches for the handle. She tries to turn it but realizes she will be too late. She turns and faces the dog just as it leaps on her.

DISSOLVE

INT HOSPITAL—DAY

MONTAGE—Starting on a close-up of the installer's wincing face, CAMERA SLOWLY DOLLYS BACK to REVEAL that she is lying on a hospital bed getting first a shot, and then stitches in her arm. Later, it's her leg and, still later, her hand. At various times during the montage, we see that she is in a good deal of pain.

DISSOLVE

If we apply our three "life speeds" to these scenes, we can see that the first one would probably have a relaxed but typical pace. The director's coverage of the scene would probably be shot with typical focal lengths and camera perspectives. It would later be cut together at a typically relaxed pace also.

The second scene would probably start typically enough, but as the dog suddenly appeared, the pace could accelerate rapidly. This would be accomplished partly in the edit bay, but during preproduction and production the director might consider shooting more close-ups—for instance, of the installer's facial reactions during the attack, the installer's hand on the gate latch, and the dog's feet as it leaves the ground leaping toward the installer.

Later, this type of footage would be cut together much more rapidly and in much shorter pieces than the footage for the previous scene. This would create a much faster "life speed."

In the third scene, the montage would tend to create a slower "life speed." The purpose of a montage is to compress time. In this case, it would leave the visual suggestion that the painful shots and stitches were "taking forever."

Just as in dramatic scenes, the pace and timing of humorous material is also extremely important. Funny lines delivered a split second too late can lose all their impact, as can a misdelivered glance, gesture, or subtle inflection.

Simple host-type scenes should normally have a typical pace, since they are meant to give the illusion of live conversational situations.

Again, the director arrives at a sense of how her program should be paced through careful script analysis, with special attention to shooting and editing considerations. Should she shoot long pieces of action that require no editing or short pieces? Will the camera move or be stationary? What is the tone or mood to be established?

Summary

As you can see, analyzing the script is the director's first and one of her most important tasks. Attempting to direct a script without careful analysis of all the aspects we have discussed is analogous to trying to play a piece of music without understanding how the notes on the page interrelate to create rhythms, tempos, moods, and varying paces.

After careful script analysis, however, the director gains a precise vision of what she must accomplish with each character, scene, prop, and location. She is thus well prepared to move into preproduction with confidence that she is guiding production of the script toward a successful finish.

6 Human Aesthetics

If a film or television program is skillfully written, carefully prepared down to the tiniest detail, and lit and shot with great professionalism, but its actors lack credibility, viewers will "turn off" to it at once. By contrast, a talented cast performing in a program prepared and shot by mediocre standards may be not only acceptable but *gripping* to its audience.

This is because, as we have noted, actors are the human part of the director's illusion. They are what we as viewers identify with most and, as such, are, without question, the most critical elements on the screen. It stands to reason, then, that the most important tasks a director performs are those of human aesthetics.

Image and Performance—The Critical Ingredients

Human aesthetics are those pertaining to actors' *images* and *performances.* To create a convincing performance and thus sustain the director's illusion, both of these qualities must appear to be natural characteristics of a scene and the actors within it.

One of the director's primary roles is to see that this true-to-life quality is present before she "buys" or accepts any scene being recorded. With professional actors—or "talent," as they are often called—she accomplishes this by paying great attention to four critical areas of human aesthetics: *casting, rehearsal, execution,* and *communication.* With *non*-professional talent, such as company employees, her concerns are basically the same but with somewhat different emphasis.

As we now cover the first three of these key areas, keep in mind that we will examine the fourth, the director's communications skills, in much greater depth in Chapter 10.

Casting

Casting is the process of contacting, auditioning, selecting, and hiring actors and actresses. Its purpose is to assemble a cast that fits the director's vision of each role based on her previous script analysis. Like every other aspect of human aesthetics, proper casting is essential to a successful program.

Auditions

In feature films and commercial television productions, auditions are often an extensive process involving weeks of interviews and readings, hundreds of actors, and multiple callback sessions. These "casting calls" are often set up and administered by casting services and casting directors who do much of the initial weeding-out process before the film or program director ever gets involved.

In corporate television, casting is handled on a much smaller scale. Actor's selections are usually accomplished based on the previous experiences of the producer or director or a 1- or 2-day audition in which perhaps 10 to 40 actors are seen.

A corporate director may have only a host role to fill that requires no unusual look or physical characteristics. In this case, she may cast the role with a person with whom she has recently worked, or based on a request by the producer or sometimes even the client. If her program has multiple roles, however, or if those roles require special characteristics or particular "looks," an audition is usually in order.

Typically, the audition is set up by someone like the AD during the initial phases of preproduction. The AD calls casting agencies and views previously produced programs in which the director may remember some actor who impressed her. The AD also looks over photographic "composites" and "head shots" kept on file. From these sources, he develops a list of potential actors and some visual representation of each. He and the director review the list and decide which actors to call in for "readings." Many times, auditions are videotaped, allowing the director to review the performances again later, when making her final decisions.

During auditions, the director carefully scrutinizes four basic qualities in each actor she sees: *appearance (look), performance capability, ability to take direction,* and *personality.*

Appearance ("look") In most corporate programs, the actor's "look," as it is often called, is important in a somewhat paradoxical sense. That is, it should appear so natural to the role that it does *not* call attention to itself. If the script calls for a gymnastics instructor, for instance, a lean, well-built individual could do the role looking perfectly natural in a gym environment. If such a role were cast with an overweight individual, however, the audience would immediately notice the actor's physical makeup as being odd or unnatural.

Many times, a look need not be physically specific in corporate programs. For instance, a role may call for someone like Bill Moller, in this scene:

INT LOBBY—DAY
John walks around the corner and finds BILL MOLLER, a communications systems sales rep, spreading his brochures out on the desk.

If Bill Moller's role is a minor one, he could be tall or short, overweight or thin, and the audience probably wouldn't notice much one way or the other. In this case, his wardrobe and makeup would be more important. If he were clean-shaven in a suit and tie he would appear natural as a sales rep. Three days' worth of stubble, on the other hand, and a pair of sweats or Levi's would look noticeably unnatural.

An actor's look may also be determined based to some degree on the audience makeup of the program. For instance, a program on cosmetic application would probably be shown mostly to women. A female host would identify best with and seem most natural to the audience. A program on backhoe operation, on the other hand, would probably be shown to mostly male construction workers; thus, it should be hosted by a male.

Following her careful script analysis, a director should have a keen sense of the types of looks she is after before her auditions begin.

Performance capability While an actor's look is important, his ability to perform is the most critical element considered in an audition. Unfortunately, the time constraints and logistics of most corporate auditions are often not the most conducive to quality performances.

Facilities are typically small and makeshift—perhaps a conference room or corner desk; actors get only a short time to review the script pages to be performed—often 5 minutes; and the time available for the actor and director to spend together is usually only 5 to 10 minutes. Add to this the fact that actors are often "starving" for work and sensitive to rejection and you can imagine why auditions can be extremely tense interactions for them.

The director, however, does not want tension. She wants a relaxed, comfortable atmosphere in which the actors can deliver their best under the circumstances. In order to achieve this atmosphere and encourage solid performances, a director can take a number of steps:

1. **Give the actors minimum material and maximum time.** If the amount of material an actor is given is kept to a minimum (a well-chosen single page is often enough) and as much time as the director can allow is provided, a better performance will usually result.
2. **Discuss the role before the reading.** An actor given a page or two of a script will have no sense of the "big picture" that might give her a better grasp of the role she will be asked to perform. A quick overview covering what the program is about and how her role fits into that big picture is often a good enough framework to prompt a more accurate and comfortable reading.
3. **Talk about any specifics you may be looking for in the performance.** If some character nuance you are hoping to see is not apparent in the script, make the actor aware of it. For instance, if you personally feel a character should be more animated or nervous than the script indicates, you should mention this as you discuss the role.
4. **Give the actor someone to perform with—when possible, another actor.** Actors work best when they have someone else to play off who has a good sense of timing and delivery. For instance, a scene in which an actor must continually interrupt another person would be difficult to

do when played off a person with no sense of when to start and stop speaking. One way to get this type of interplay is by auditioning two or more actors at a time, letting each work different roles and perhaps even switching roles midway through the audition.

5. **Never play the other part.** A director who plays the opposite role cannot hope to accurately judge an actor's performance—at least, not during the initial audition. If, due to budget constraints, you must read opposite an actor, be sure to videotape the auditions and save your judgments until you can view the tapes uninterrupted.

6. **Give direction and a second chance.** If an actor does not deliver what you're after on the first read, unless you happen to be in an extreme time crunch or unless it is painfully obvious he or she couldn't hope to get it right, offer direction and give the person at least one more opportunity. Sometimes a little encouragement and a second try can work wonders.

Although auditions can probably never be truly comfortable meetings between directors and actors, the director who follows these steps can at least make them as comfortable and productive as possible.

Ability to take direction The ability to take direction means the ability to change a performance based on the director's wishes. A director can quickly find out how well an actor takes direction in an audition by asking her to change her delivery.

For instance, if an actor is doing a role as a customer service representative on the telephone and her initial direction was to *please* the customer on the other end, the director might follow her performance by saying something like, "That was very nice. Now let's approach it a little differently. Instead of pleasing the customer, let's do the same lines but try to *insult* her."

As you can guess, this requires a completely different treatment of the same words. Tone, inflection, pace, gestures, and facial expressions would all change.

If the actor does this well but the director is still not convinced, she should follow by asking for some *degree* of change. For instance, "Great. That would insult anyone. Nice job. Can we try it once more, though, and take it only *half* that far? I wouldn't want this customer to be given a reason to hang up and call the boss; I'd just want her to leave the conversation with a bad taste in her mouth." If the actor can deliver again, the director can feel comfortable she is able to take direction quite well.

Personality A director and her production crew must sometimes work closely with an actor for days. This means eating together, taking breaks together, traveling together, collaborating on the set or location to create credible performances, and, most of all, maintaining a productive working relationship through difficult and at times extremely pressured situations.

If, during an audition, an actor appears to be the type that might be difficult or overly demanding, it's a good bet he'll be even more difficult in a production situation—when lunch is late, rain is on the way, and the shoot is an hour behind schedule.

In feature films, the needs of actors and actresses are often seen to by staffs of people whose jobs are simply to keep the actors content and ready to perform when their scenes are shot. These people provide everything from massages and portable Jacuzzis to decks of cards and shoulders to cry on.

In corporate television, these luxuries are rarely the case. In many instances, besides having no frills whatsoever, actors must apply their own makeup and provide their own wardrobe.

All these factors point up the need for a director not only to be sure that her actors are able to perform and take direction but also to remain positive and helpful under the sometimes tense situations that are part of the television production process.

Audition Review

When the auditions are finally over, the director will review the performances of those actors she is strongly considering. As she does this, she will either consciously or unconsciously refer back to her script analysis. What should the person in the script look like? How should he move in front of the camera? What tone of voice, facial expression, or gestures should he use? Which actor is best able to do those things most convincingly? When her decisions are made and the actors are booked, the next step in the human aesthetic process is rehearsal.

Rehearsal

Rehearsal in network television means gathering the actors together at the studio in the days before production. Working with the director, the actors read, revise, and eventually "block" the script, working out entrances, exits, and other movements in front of the cameras.

In high-budget feature films, the same process is carried out, but often on a far more elaborate scale. The cast may assemble in Spain, Japan, or whatever other country the film happens to be shooting in. Actors visit locations weeks or months in advance to get a "feel" for them. They may also study local history and culture as part of their research. When rehearsals take place, they might happen on the bank of a river, in the tower of a castle, or on top of a mountain—wherever those scenes are to be shot.

In corporate television, a rehearsal may not happen on a soundstage or in a castle tower, but some degree of rehearsal should always take place. The reason is simple: Even a short period of rehearsal between only the actors and director can save considerable time on the set or location while a full production crew stands by looking on. Depending on the budget and time available and the size and type of the cast, rehearsals can take place either before or during a corporate shoot.

Before the Shoot

Full advance talent rehearsals are rare in corporate television. This is the case for two primary reasons. First, there is rarely enough time or money available for full rehearsals. Second, most times the roles are simple enough to negate the necessity for a rehearsal. In many cases, however, some degree of talent rehearsals does get scheduled. When this is the case, the director usually works in three steps: *script discussions, readings,* and *blocking.*

Script discussions Simple script discussions between the director and her cast often take place over lunches or sitting around conference tables. Topics often include character traits, motivation, difficult dialogue sections, and plot elements. Informal as they may be, these types of meetings often clear up a good many minor questions that would later have proven costly.

For instance, an actor's interpretation of how a character should display a certain emotion may differ from the director's. If these differences aren't discussed and resolved in advance, they will have to be worked out on the set. On the set, however, they will become apparent only after at least one unsatisfactory run-through, and the ensuing discussion will hold up a full crew of perhaps five to ten people while the actor and director reach an accord.

Readings If the director is allowed more time than is required for an informal script discussion, her next step would be to have the cast read through the script several times. This allows the actors and director not only to talk about how performances might vary but also to actually test out those variations.

In addition, rough spots in the script are often discovered in read-throughs. This means that on-the-spot revisions can often take place before they become stopping points on the set or location.

Blocking The ideal, though rare, situation for a director is to have rehearsal time set aside with her cast on the set or location on which they will work. When this is the case, the entire performance, including any nuances, rough spots, and even blocking, can be worked out. When the actual set or location is not available, blocking can still be worked out to a large degree by using prop furniture and similar rooms or areas.

As the actors read, they also walk through each scene and get a feel for things like when to turn, when and where to walk, and how to gesture and glance. The director watches such a rehearsal and stops it with comments like: "The way you turned on the spot and faced him was powerful. It worked well. But I don't think it's time yet to walk up to him. In a few minutes, you're much more upset. That would seem to be the time to move toward him. Let's try holding that move till page 5." In cases like this, since it may be impractical to go through the entire script with all the actors, scene by scene, a common practice is to rehearse and block the most complicated scenes or those which will require the most time to get right in production.

If the director is not given time to meet with her actors face to face, she should still insist on some small amount of rehearsal preparation for her cast. She can accomplish this in a number of ways.

Call at home Most actors and actresses are delighted to receive an advance call from the director to discuss a particular role. While a face-to-face discussion with the rest of the cast is obviously preferable, a phone conversation can still clear up *lots* of little questions.

Mail or fax If a telephone conversation won't do, try corresponding by mail or fax. At the very least, a director can write up a short overview of how she envisions the character to be played, nuances or special instructions, and perhaps a general interpretation of the role as it relates to the other characters or the overall script. Fax or mail correspondence can sometimes be carried out via the actor's agent.

Rehearsal on the Set

As we have noted, the amount of advance rehearsal a corporate director gets is, in most cases, minimal. In these instances, it becomes necessary to rehearse on the set just prior to each take. When this is the case, a few hints can help streamline the process and make it most effective.

Give an initial scene overview It always helps to be given a framework within which the performance should take place. For instance, imagine that the scene to be played is the "open" of what the director knows is a very positive program. The action involves a host entrance, delivery to camera, and an exit.

The director might tell the actor something like the following: "This is the open, and I'd like to set a positive tone right away, so we'll try an entrance from over here; a nice, conversational, upbeat tone; and when you get to the lines 'Let's find out,' just turn and exit to your left. We'll be starting in a wide shot to establish you and the set, but we'll zoom in after a few seconds. So when you make your exit, we'll have you in a medium close-up—about chest up."

This type of quick overview would set the stage nicely. The director should next elicit any questions that might be on the actor's mind. Once these were taken care of, a rehearsal in front of the camera would follow.

If the scene is a dramatic one, involving, say, two role-play actors having an argument, the director should still offer her sense of the scene in overview fashion, and she should still have suggested blocking. Final blocking for dramatic scenes, however, often must be worked out as the actors actually perform the scene.

In such a case, the director might say, "So this is the scene where Andrea blows her top. My sense of what's happening here is that since she's been holding it in for months, now she's ready to just punish William with everything she's got. I think we'll start with a master shot with the camera over there. I thought it would work out naturally if you're just turning back from the window, Dale, as she rushes in the door. That way, you're both suddenly thrust up face to face. What do you think?"

At this point, the actors might have questions or suggestions. The director should listen to these, explore the reasoning behind them, and reach a consensus on what to include or not. When an agreement had been reached, the scene could be rehearsed.

Be specific by offering examples or analogies When an actor is performing in a way the director does not like, it's very important that the director clearly identifies what's bothering her and why. Both she and the actor can then set about changing the performance as needed.

For instance, in the scene mentioned above, if the Andrea character were being overplayed, as the rehearsal finished, the director should *not* say something vague and confusing like, "Good. I like that. The emotional content is great. That's what I'm not positive about. The level is pretty intense. I mean, it's intensity I'm after, but, well, I'm not sure about the content and the level of that intensity."

Instead, she might say, "Very nice. Well done. But I'm a little uncomfortable with the intensity you're giving it, Joan. My sense is that Andrea would feel a more controlled, seething kind of anger. What do you think?"

If the actor disagreed or had questions, more discussion would follow. If the disagreement remained, the director should be democratic to the extent she is able. In the end, however, she is the one responsible for what goes on the screen, and she should make the final decisions. Actors realize this and are generally willing to go along with the director's wishes.

Don't wait until later to give specific changes When a performance requires changes, the director should not wait for four or five takes before she communicates these adjustments. As actors go through a scene over and over, they become more and more attuned to its flow and pace and certain key events or pieces of dialogue. They may mentally "connect" a turn of the head with a certain line, or perhaps tie a gesture to a specific inflection.

The best directors make sure their criticism is heard *before* this ingraining process makes it increasingly difficult for the actor to change. As with any rule, however, there are always exceptions. In this case, one exception is when a scene is rough in general and the director feels that a few more rehearsals will smooth it out. Another exception is when an actor delivers some tone or inflection that the director feels was just a quirk on that particular take. In these cases, the director might simply say, "OK, a little rough around the edges, but I think we just need to warm up to it. Let's try it again."

Provide encouragement Acting is an extremely vulnerable profession. Actors are often very sensitive and keenly perceptive individuals who are sometimes hurt by rejection and virtually always motivated by recognition.

With this in mind, the director should always provide immediate recognition and encouragement to the members of her cast when these are deserved. Such praise can come in the form of a sincere but brief comment like "Very nice," "Well done," or "Convinced me!" With this type of professional "applause," an actor will be eager to please and much more likely to accept constructive criticism when it's offered.

What the director should *not* do, however, is patronize an actor or overplay his sense of satisfaction. As we have said, actors are also perceptive people. They are quick to see through and lose respect for "syrupy" directors who are obviously manipulative and insincere.

Be open to suggestions. Often the director of photography, assistant director, or perhaps a grip will catch things you may miss.

—Ralph Phillips
Writer/Director

Break for discussion if needed If your scene is simply not working, it's best to break and take 5 minutes to think it over or discuss it with the actors. It can also help in these cases to ask the AD or another trusted crew member for an outside perspective. At times, even the best directors and actors can get too close to their material, and an objective point of view can provide a new sense of focus or perspective.

It also helps during these times to focus on the word *motivation.* A lack of true motivation is often what makes a scene seem unnatural or uncomfortable.

To alleviate any improperly motivated material, the director should go over the scene carefully in her mind, analyzing each line and movement and asking, "Is this called for?" or "Would this character really do or say *this,* in this situation?"

Be decisive Finally, remember that a wishy-washy director usually gets little respect from either her crew or her actors. A director with a clear sense of vision, on the other hand, is prepared to provide the leadership and direction that both the crew and the cast require.

Even more admirable to the crew is a director who knows what she wants but is willing to let the crew and actors have input into how her vision is to be accomplished.

Execution

If a director's casting and rehearsal have been productive, the actual execution of the scene on the set or location is normally as simple as saying "Action" and "Cut" several times while waiting for the perfect take to occur. By this time, the rough spots have been uncovered and dealt with, the actors and director are both clearly focused on what they must accomplish, and the crew is in place and ready to do its part.

Distractions

The most frequent problems during a well-rehearsed execution come from irritating distractions or frustrating circumstances at times beyond the control of the director. For instance, if a scene being played out on location is going extremely well but three times in a row a plane flies overhead, ruining the take, both the director and the actors will become frustrated.

Another irritating distraction that sometimes hampers corporate video shoots is equipment absence or failure. Nothing is more frustrating than to have a performance fine-tuned and ready to record, only to have a camera break down or a generator run out of gas. The negative impact these types of problems can produce is multiplied as other complications, such as time constraints, also become factors.

Many of these types of distractions can be alleviated in preproduction. Careful choices of locations and thorough planning and organization before the actual shoot days can make the production days much more enjoyable and productive.

A director's golden rule, however, whether or not distractions occur, is to remain calm, confident, and in control. As previously mentioned, crew and cast members look to the director for leadership. When frustrating times occur, if the director tends to lose her temper or composure, it will create a negative tone on the shoot that could seriously affect its outcome.

Directing Nonprofessional Talent

In high-budget feature film and commercial work, a director has the luxury of working with consistently professional and often highly paid talent. In corporate television, however, due to budget restrictions, this is often not the case.

Although most corporate programs rely on professional actors to do key roles, such as hosts and principal characters, there are still a good many times when the corporate director must work with company employees. This is usually the case when shooting employee interviews, executive messages, and employees involved in some form of work activity.

Under some circumstances, this type of footage can work quite well. In other cases it can be disastrous to the credibility of the program. A director should keep in mind several basic rules that will greatly improve her chances of success when working with nonprofessional talent.

Do Use Employees to Demonstrate

The most credible roles employees can play in corporate programs are those of people like themselves. A telephone installer, for instance, stands a good chance of doing a credible job in a scene that calls for someone to greet a customer and repair her phone. A company secretary would be well equipped to fill the role of a clerical employee going through, perhaps, the four steps involved in proper records retention. Likewise, a warehouse worker could probably do a convincing job of demonstrating how to stock shelves safely.

In each of these cases, the key is the fact that the employee is *demonstrating* rather than acting. Since these types of demonstrative scenes are common in corporate programs, there are a good many instances when employees can fill them successfully.

Two primary cautions apply, however, when using employees even in this limited "talent" capacity: *over directing* and *giving employees scripted dialogue.*

Over directing Over directing means requiring an employee to do anything more than simply be herself and demonstrate the job at hand. Some examples are asking the employee to hit the DP's camera mark on an entrance or exit, telling her to lean or turn a certain way to accommodate lighting or framing, encouraging her to gesture, glance, or motion in some way that is not normally part of the job activity being performed. In each of these cases the result is almost always a stilted, obviously staged performance.

Stilted dialogue When an employee is required to speak dialogue, she is removed immediately from the demonstrative role and placed in a character one. As we will discuss shortly, virtually all attempts to make this work will fail. One compromise, however, is to use an employee's dialogue as audio *under* a host's narration. When used in this way, although the dialogue is present, stilted or not it simply provides a background "bed" of audio ambience under the host's predominant voice.

Do Not Cast Employees in Character Roles

Acting is a highly complicated skill. It seems amazing to me that a client would invest thousands of dollars in a company video program but allow—and, in fact, sometimes even request or *insist*—that one of the most critical elements in that program be left to people totally unprepared to deliver what will be expected of them.

This is the case when employees are given character roles. The reason is obvious. It takes years of hard work, study, and dedication to become a credible actor. Employees have none of this background. They are totally unfamiliar with the production process and in addition are often dazzled by the "glamour" of lights, camera, and so on. As a result, when they are put in this position, what normally ends up happening is this:

Laughable performances are the best that can be recorded.
The employee is embarrassed.
The program loses all credibility—*if* the material ever gets used.
The client is left unhappy, and the director's and producer's work looks terribly unprofessional.

Hopefully the producers the director works with will have explained this to the clients and she will never be placed in such a position.

Do Not Give Employees Scripted Information Lines

This sometimes happens when an employee is to be interviewed or must look directly at the camera and get specific information across.

When that information is being delivered by an executive such as the CEO, scripted information lines become an unavoidable necessity. Such messages must contain only certain points, often stated in specific ways. Thus, they are prepared in advance. The talking-head CEO, however, is forgivable to a large degree because employees *expect* such presentations and realize that his remarks require the advance preparation just mentioned.

On the other hand, when the information is being delivered by a regular employee who is supposed to be sincerely making a point or explaining some job function, the result is usually a complete loss of credibility. The reason is the same one mentioned earlier. Employees are *not actors,* and scripted lines delivered to either a camera, as information, or another person require solid acting skills.

The answer to this dilemma is often what's called the news-style interview.

Do Employ the News-style Interview

In the news-style interview, the employee is asked to face just away from the camera, looking at an off-camera interviewer—normally, the director. A series of ideas or topics are suggested to the employee, in the form of either questions or statements, and she articulates her answer in her own way. Later, the material is edited to remove redundancies, pauses, stumbles, or extraneous comments. News-style interviews allow employees to maintain their dignity and composure while still communicating important information. Just as important to the corporate director, the interview format also adds a good deal of credibility to most programs.

Try to Meet with Employees Before Taping Interviews

This is sometimes not possible on corporate shoots, due to time or budget constraints. Most times, however, it *can* be arranged, and, when the luxury is extended, the director should take advantage of it.

When conducting such meetings, the director should make the employee feel as comfortable as possible and simply talk about the subject of the video interview. As the discussion gets under way, the employee's *look* and *ability to articulate* ideas are the critical considerations.

If both these elements appear positive—especially the ability to articulate and express ideas, the on-camera interview will probably go well. Keep in mind, however, that even after a successful preproduction interview, some employees freeze up when the lights and camera actually arrive.

Play Down the "Hollywood" Image

One surefire way to make an employee nervous is to play up the fact that she is about to "become a star." Employee segments or interviews are best handled on a business level, and employees should be encouraged to forget the lights, crew, and camera as much as is humanly possible.

Be an Involved Listener

If an employee being interviewed feels the director is really listening, she will tend to forget about the situation at hand and become involved in making credible points.

The director can be an involved listener by, first, sincerely caring about the subject at hand and the employee's opinions or ideas. Second, the director can encourage the employee to continue speaking by nodding often or saying things like, "I see," "Yes, I understand," "Right, right," or "Now I get it. So tell me more about. . . ." Third, the director should always ask open-ended questions. These are questions that cannot be answered with a simple yes or no; they encourage thought and discussion rather than single statements.

A few examples of both open- and closed-ended questions are the following:

Open-ended questions Tell me, how would this happen? I'd like to hear more about that? What kinds of steps are involved?

Closed-ended questions Would it happen like that? You would say that's good, I assume? There are three steps involved, right?

Prepare Interview Questions in Advance

A good way to keep a discussion moving while getting at the key issues involved is to carefully think out and write down the questions employees will be asked. As a part of this preparation, the director should first

To reduce the anxiety of nonprofessional talent, let them see themselves on a monitor before shooting. Then remove the monitor before rolling tape.

—Dan F. Johnson, M.A.
Producer/Director

consider not the question but the *responses* he hopes to elicit. The questions should then be determined accordingly.

For instance, if an employee is to be interviewed about a new work policy and the director hopes to hear that the policy is well accepted and was easy to implement, he might ask:

> Tell me about the new policy—what does it entail?
> Just how does it affect you and the other people on the job?
> Is it a good policy? If it is, tell me, for what reasons?
> Overall, tell me, why will the policy and its implementation be a positive change for your department?
> How about your own personal feelings, or anything else you'd like to add?

These questions cover the areas of concern several times from slightly different angles. They would thus give the employee plenty of chances to find positive things to say about the new policy. Asking for a personal opinion as a last question often elicits good ideas that may previously have slipped the employee's mind.

Be Patient

Last but not least, be prepared to let the tape roll, if necessary, or wait for the employee to relax. Many poor interview pieces could have been considerably improved simply by allowing the employee a chance to become oriented enough to the surrounding crew and equipment to begin to *forget* them.

The director can help those processes by assuring employees that there is plenty of time and encouraging them to gather their thoughts before speaking. I usually follow such a statement by saying something like this:

> To be frank with you, when we do these interviews, we use only a few seconds of what the person says. You may talk for 10 minutes, but we'll maybe end up using 30 seconds or so. Oh, and by the way, we can edit out any stumbles or mumbles, so don't worry about getting it all out in one breath. Just take your time, think about it, and say it however you like.

So . . .

In the final analysis, a director's ability to shape the human aesthetics on her production are directly related to her ability to communicate with, empathize with, and guide the efforts of her cast, crew, and the company employees she works with.

As mentioned, we will discuss the communications aspects of the director's work much more closely in Chapter 10. For the moment, however, let's move on to audiovisual aesthetics and examine their relationship to the directing process.

Audiovisual Aesthetics

7

The director has now carefully analyzed the script and, based on this analysis, formulated character roles. He has also begun the delicate process of guiding his cast toward credible performances. The elements he has executed thus far could be thought of as one side of a director's aesthetic balance.

Now comes the other side of that balance: the task of making sure those performances are recorded on videotape in a way that will not only maintain but also heighten their impact and credibility. The first element on this second side of the balance is audiovisual aesthetics.

Pictures and Sound

Audiovisual aesthetic qualities are those pertaining to the audio (sound) and video (picture) elements of each scene. They are elements such as lighting, focal lengths, frame compositions, camera movement and placement, and sound recording. The use of audiovisual aesthetic elements often changes depending on the type of program being produced.

As an example, the audiovisual elements in a simple "host on camera" program might include an office set, lighting with little contrast or modeling, typical midrange focal lengths, and camera positions such as eye-level medium shots. It would probably also include standard sound recording systems and techniques.

In a dramatic scene, however, the audiovisual elements would change. For instance, the set might be something like a darkened room. If so, dramatic, high-contrast lighting would probably be used; more extreme, close-up focal lengths might be incorporated; and perhaps nonstandard camera angles, such as overhead or low angles, would be employed as well.

When these types of audiovisual elements are shaped and guided properly and added to the performances the cast puts forth, the two separate sides of the director's balance blend as one solid, audiovisual entity—a shot, scene, or entire program that is completely credible to its audience.

The director's job involves balancing the technical (light levels, mic placement, camera moves, and so on) with the creative (visual composition, color, texture, and emotion).

—Patti Ryan
Producer/Director

Script Analysis and Visualization

Just as with human aesthetics, audiovisual aesthetics are at first based on script analysis. As the director reads and visualizes each scene, he begins to "see" the audiovisual elements it is made of: angles, lighting, locations, and so on. This visualization process continues throughout the preproduction phase of the director's work. As production nears, his visualization gains focus and detail. Finally, during production, the audiovisual aesthetic qualities of every scene must be shaped, guided, and judged acceptable by the director on each take.

Audiovisual aesthetics fall into two general categories: *visual elements* and *sound elements.*

Visuals

Visuals mean pictures, and when it comes to pictures, the director of photography or DP is the director's closest collaborator. It is the DP's job to help bring the director's vision to the screen with photographic—or, in the case of video, videographic—impact and credibility.

In high-budget feature films, the DP typically does not operate the camera. His main functions are to plan, create, and supervise the lighting of each set or scene and guide the camera crews in such areas as the use of appropriate film stock, camera positions and movement, and filter and lens usage. The DP often has a full staff at his disposal for these purposes, and their single overriding objective is to accomplish what the director has asked for in terms of the mood and look of his pictures.

In corporate television, the DP (sometimes called the videographer) usually handles lighting *and* camera operation. He may have one or two people, perhaps a grip and a gaffer, to help place cables and lights and to move the camera and lighting equipment. On many smaller corporate shoots, the DP is on his own—setting up lights and camera by himself.

Though the scope of his job may be smaller than that in high-budget features, his objective remains the same: to give the director the visual aesthetic quality he is after. The DP accomplishes this task by orchestrating four elements under the director's supervision: *camera placement and movement, frame size, frame composition,* and *lighting.*

Although the DP or his staff may be physically executing these elements, the director is the overall guiding force and the final decision-maker behind them. For this reason, he must have a complete understanding of the DP's role and responsibilities. He must also know how the elements just mentioned affect the audiovisual aesthetics of his program. In order to gain that understanding, let's examine each of these critical areas in greater detail.

Camera Placement and Movement

When considering camera placement, the director should first and foremost remember that it is the perspective he chooses to give his audience. This perspective can be singular, as in the case of one camera shot that records an entire scene, or it can have multiple perspectives, as in the case of an event recorded from several angles that is later edited into one scene. It can also be a stationary or moving perspective.

In order to arrive at a decision on camera placement and movement, the director once again reflects on his script analysis, the blocking or movement he has worked out with his actors during rehearsal, and the resources available to him. With all these issues considered, he must ask himself, "What is the best place or *places* for the audience to view this action? Should the camera be at eye level? Should it be fixed in terms of position or focal length? Should it follow the action by actually moving along with it, or simply pan left or right as the action takes place?"

As you might guess, many technical ramifications are involved in this decision. Just a few might be these: What type of camera equipment is available? Is there a dolly? What type of lenses are being used—a zoom or fixed focal lengths? How much time is available on the schedule? Is the camera operator capable? Is the talent capable?

Standard coverage The simplest and safest way to record a scene is to shoot the identical action from a number of typically accepted camera angles and focal lengths. This is referred to as shooting "standard coverage."

Standard coverage calls for the director to place the camera most times at eye level and first record a *master scene.* This is usually a wide angle, encompassing all action in the scene and all actors.

After a satisfactory take or "buy", as it's called, of this master, the camera is moved in for *medium shots* or *medium close-ups.* These are much closer angles that encompass each individual actor's part in the scene. They are recorded with each actor repeating lines and action identical with that of the master scene. When a "buy" of each of these is recorded, *close-ups* of certain parts or *inserts* are recorded in the same way. These are still closer angles that encompass very small parts of the same action.

The result is a scene recorded identically from a number of angles. Later in the editing process, this coverage allows the director to provide the viewer with a series of effective and aesthetically pleasing perspectives as the scene unfolds.

To illustrate, let's use a very simple example.

INT SERVICE OFFICE—DAY
JEAN, a customer rep, is talking to a customer on the phone.

JEAN
Yes, ma'am, we do provide a warranty for the first 2 years . . .

CAROL, her supervisor, enters and sits down. Jean can see that Carol wants to say something important.

JEAN
Excuse me, ma'am. Could you hold for one second, please? . . . Thank you.

She places the customer on hold.

CAROL
Sorry to interrupt, Jean. I just need to know right away if you're signed up for the first training session. The cards have to go by three today.

JEAN
Yes, I did sign up. Donna has my card.

CAROL
Great. I'll get it off, then, with the rest of ours.

Carol now gets up and exits. Jean releases her hold button and continues with her customer.

JEAN
I'm back, ma'am. Thank you for waiting. Now, about that warranty, you can sign up for it by phone . . .

DISSOLVE

As shown in Figure 7.1, we could shoot standard coverage of this scene as follows:

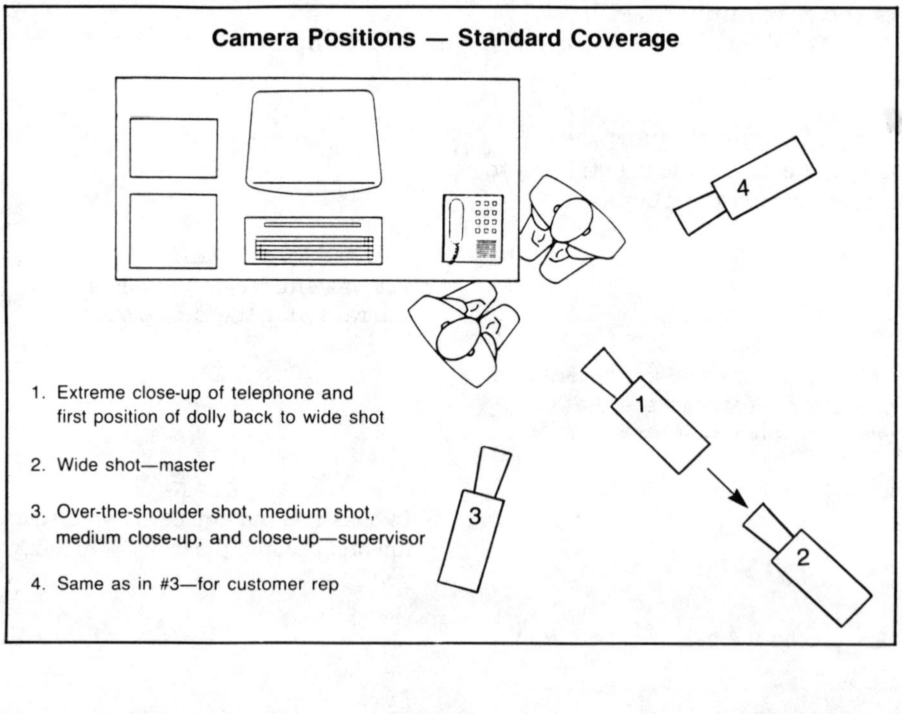

Figure 7.1 Camera positions for a typical corporate setup: supervisor on right and customer rep on left. Camera angles 1 through 4 will provide standard coverage.

The camera is positioned at approximately eye level for a seated person. It is mounted on a dolly and rolled in to a point where it can begin the shot by giving the viewer a close-up of the telephone the representative is talking on.

The *master shot* is recorded beginning with this close-up of the phone. As the representative starts her dialogue with the customer, the camera is rolled (dollied) back to a position in which both the representative at her desk and a second chair beside her desk are visible. A small part of the surrounding office can also be seen. As this "dolly back" is nearing its end point, the supervisor enters the scene and takes a seat. She and the representative talk, after the representative puts the customer on hold. The supervisor finishes and leaves, and the representative releases the hold button and begins talking again with the customer.

This initial angle would provide a profile view of both actresses during the entire scene, as well as an "establishing" perspective of their immediate surroundings—a service office. The shot would begin, however, with a close-up of the phone that might allow for a comfortable transition from the previous shot—perhaps a host on camera.

Once a "buy" of this master had been recorded, the director might ask that the camera be kept at the same level but moved to a position looking over the supervisor's shoulder at a *medium shot* of the representative. This would allow the viewer to see the representative completely and some of the desk and computer in front of her. The scene would then be acted out again, identically as it had in the master shot. As the representative began to talk, we would see the back of the supervisor move in from camera left to right as she entered the shot. She would take a seat, remain there with her back to us as the scene played out, and move out of the frame when the conversation had ended. Again, the representative would hit the hold button a second time and continue her conversation with the customer.

The director might now ask that the camera stay at the same or a close by position and be zoomed in to a *medium close-up* shot in which only the face and upper body of the representative were visible. The scene would again be played out and a "buy" recorded.

The camera would then be moved to the other side of the desk, looking over the representative's shoulder at the supervisor. It would remain in basically the same focal length as the previous over-the-shoulder angle—a *medium shot.* The scene would again play out completely and identically to the master until a buy had been recorded. The camera would again be moved in to a close-up—this time, of the supervisor—and the scene would be recorded a final time.

Finally, the director would ask that the camera be moved to a position where it could provide an extreme close-up of the hold button on the telephone. From this position, the director would record the operator's finger entering the frame and pushing in the hold button. He would then record a second version, this time of the representative's finger entering the frame and releasing the button.

We should note that in an actual shooting situation, these final inserts might have been shot in the very beginning, when the camera had been dollied in for the opening close-up of the master shot. We've placed this piece of the coverage last in this instance for the sake of clarity and simplicity.

Having accomplished this, the director would have recorded standard and very adequate "coverage" of the scene for use in editing.

Editing coverage Perhaps days later when the editor had finished cutting this footage together, the scene would begin with the close-up of the phone. As the camera dollied back, the wide master shot (including the representative) would be revealed. We would hold on to this shot until the supervisor entered, took a seat, and began talking. At this point, the editor would cut back and forth or "intercut" using the over-the-shoulder and close-up shots. This would allow us to see the supervisor and the representative individually as each one spoke. When the rep put the customer on hold and later released her, the editor might briefly use the extreme close-ups of the rep's finger to show us that action taking place. Finally, he would probably cut back to the master shot just as the supervisor was getting up to make her exit. This type of shooting and editing would provide the audience with an establishing sense of where the scene was taking place and a series of very comfortable perspectives of all the action as it played out.

Nonstandard Camera Placement and Movement

As we've just seen, standard camera placement is usually at eye level to the actors, in a position that will provide a clear view of the action. This type of standard camera placement is appropriate in the vast majority of corporate shooting situations. In some situations, however, the scene might be enhanced by what could be called nonstandard camera placement. This is typically the case when a nonstandard view of the scene or a part of it would heighten its visual impact.

For instance:

FADE-IN

INT LIVING ROOM—DAY

Bill and Genie are face to face, arguing angrily.

GENIE

I've had it with you! You're never on time, and all you ever do is complain once you get here!

BILL

Yeah, well, I figure why should I be on time for a flake like you who never does anything right to begin with!

Genie has had it. She slaps Bill in the face.

This scene could be shot with standard camera placement and work fine. It could also be shot, however, from a *low, close* camera angle that would in effect place the viewer in a somewhat vulnerable position under and among the actors. The viewer would then become more affected by and involved in the anger and slapping action that takes place in the scene. This would add tension and a sense of reality.

The same scene shot from an *overhead* angle, on the other hand, would tend to provide a voyeuristic or omnipotent perspective, in which the viewer would feel safely removed from involvement in the action. Depending on what the director wanted to accomplish, either choice or a combination of both could be quite effective.

When nonstandard placement is used in this way, it becomes a seemingly natural perspective that goes unnoticed by the viewer. When such placements are used improperly, however, the viewer may become aware of the camera's presence—which is something the director rarely if ever wants to happen.

Besides heightening tension and dramatic impact, nonstandard angles can also be used simply for a fresh perspective or to enhance the visual aesthetics of a shot. For instance, I once shot a host in front of a building with a large, attractive, logo-type sign on it. When the camera was placed at eye level, we could not frame the shot so that the actor could be seen in conjunction with the logo—something I wanted to accomplish. After some discussion with my DP, we lowered the camera to about chest level and brought it in closer to the actor. This positioned us looking up at him. The shot then framed very attractively with the logo sign just over his shoulder. On another project, I was once scouting to shoot a morning meeting between a supervisor and roughly ten of his employees standing at the rear of the supervisor's truck. From eye level, I knew the shot would be acceptable, but it would not give a sense of the circular, almost powwow-type arrangement in which the men would gather. I found a nearby second-story balcony that I knew would provide more attractive perspective. That's where we placed the camera during production, and the shot worked quite well.

Movement Remember, it is the director's intent to never let the audience become aware that a camera is present in the midst of this action. When he is successful at this, the full illusory quality of a scene keeps its integrity and the audience never questions how or why they are able to eavesdrop on the events.

Camera movements, including zooms, can ruin this illusory quality, if not executed properly and with motivation. This is because unmotivated moves can make the audience aware that the camera is present and the director is "doing something" with it for dramatic effect. Keeping this in mind, remember that camera movement should be used only to emphasize, reveal, or create a needed sense of motion to enhance a shot or scene.

For instance, consider this scene:

INT KITCHEN—DAY
Marian is on the phone. As she hears the words of the police sergeant, her face registers total shock.

SERGEANT (VO)
I'm sorry to tell you this, ma'am, but
your son has been involved in a severe
motor vehicle accident. He's in critical
condition at St. Martin's hospital . . .

If the director decided to zoom or dolly rapidly in from a medium shot of Marian to a close-up as these lines were being spoken, it would certainly *emphasize* the shock Marian is feeling. At the same time, however, since the audience would, in a sense, be *sharing* that shock, they would never stop to realize the camera was moving. This is a well-motivated camera move.

Suppose Marian were having a conversation with her son, however, about what to fix him for lunch that day. If the same zoom or dolly were used as she was asking "Mustard or mayo?", it would break the illusion rather than emphasize it. This is because it would have no purpose and thus would be a totally *un*motivated move.

Let's consider another example:

FADE-IN:

INT COFFEE SHOP—DAY
A typical small trucker's diner at dawn. Bearded, baseball-capped drivers crowd a 1950s-style counter in the foreground, and roughly ten tables covered in checkered cloths fill the rest of the FRAME. DANA and MARK are seated in a far corner booth, having coffee.

DANA
How's your pickup running?

In this case, the director might choose to start this shot with a high, wide angle showing the entire interior of the restaurant. He might then have the DP pan and zoom in to the area in which Dana and Mark are seated just before Dana's line. This would first establish the setting and then *reveal* the two characters we are about to meet. Again, this is a *motivated* move and one that would enhance the shot but go unnoticed.

By contrast, assume that later in the same shot, we suddenly zoomed back *out* to the high, wide angle as Dana and Mark continue their conversation. This move would be unmotivated and might well result in an awareness by the audience that the camera was present but not covering what is should be.

For a third example, look over this scene:

FADE-IN:

INT FITNESS CENTER—DAY
An aerobics class in action. Feet hammer at the wooden floor, faces bead with sweat, and arms swing up and down in time to MUSIC. The instructor stops working out and moves among the exercisers.

INSTRUCTOR
Stay with it . . . and two and three and
four and . . . good! Knees high . . .
arms to the sky. Let's go!

While a moving camera might not reveal or emphasize any specific thing or person in this shot, a sense of motion is definitely called for. In this case, the director might want a hand-held camera to move among the exercisers or a series of rapid zooms, pans, or dollies. The general sense of motion this would create would be a natural enhancement to this type of footage—and since it would be motivated, the audience would never realize it was happening.

Camera moves can also be used to emphasize transitions. For instance:

INT OFFICE SET—DAY
The host turns to the CAMERA and says:

HOST
So the way you treat your customers can
have a definite effect on the mood you
take home—as we're about to see . . .

Since this shot is signaling a transition to what will probably be a vignette, a zoom or dolly in to a close-up—or even the reverse, an outward move—would act as a subtle cue to the audience that we are about to "leave" for some other location.

As long as these types of zooms are not overused, they work fine. If the director starts zooming or dollying on every shot, however, the transition "signal" will be replaced by an awkward, forward-and-backward sense of motion that will more than likely distract instead of enhance the scene—and probably also become apparent to the audience.

Frame Size

Frame sizes refer to the amount of area seen in a shot. Since frame sizes are directly related to the focal length of the lens being used, let's first consider this aspect of the subject.

Short focal lengths Lenses with focal lengths of approximately 10 to 25 millimeters produce wide angles and sometimes a distorted sense of distance between foreground and background objects. A wide-angle lens, for instance, can make a set piece, perhaps 5 feet behind a host, appear to be far behind him in the distance. Extreme wide-angle lenses produce fish-eye perspectives that severely distort the images. These are normally used only for special or noticeable effects.

Mid-range focal lengths Lenses with focal lengths of 30 to 50 millimeters produce images of average size and background/foreground relationships. They also produce the least amount of image distortion. They are the most commonly used focal lengths for recording master and medium shots.

Long focal lengths Lenses with focal lengths of 75 to 100 millimeters and above produce narrow angles. They also tend to magnify images and distort the distance between foreground and background objects in the opposite way wide angles do.

Long lens shots are those in which, for example, a person is seen running across a boulevard and a fast-approaching car nearly runs her over before swerving to one side at the last instant. In reality, there is a great distance between the person and the car, but the long lens tends to "squash" or "flatten" the apparent distance between the two.

Prime lenses In feature film production, an assortment of lenses called "primes" are typically rented as a part of the camera package. A set of prime lenses will typically consist of five to seven lenses, each with a set focal length ranging from short, perhaps 15 millimeters, to long, perhaps 150 millimeters. Since prime lenses contain a minimal amount of lens surfaces, they produce top-quality pictures with minimal light requirements.

Zoom lenses Zoom lenses are those with multiple focal lengths. They are adjustable by rotating a lens barrel either by hand or with the use of a small motor called a "servo." Zoom lenses are often described by the ratio between their shortest and longest focal lengths. For instance, a "10 to 1" zoom lens with a minimum focal length of 15 millimeters will have a maximum focal length of 150 millimeters.

Zoom lenses are often more handy than primes, since changing a focal length simply entails rotating a lens barrel instead of removing and replacing a lens on the camera. Feature film directors and DPs will often stick to primes, however, because zooms have more internal lenses or "glass." Each additional glass surface slightly reduces overall picture quality and increases the need for light.

In corporate television, zoom lenses are the norm and primes are the exception. This is because today's zoom lenses produce images of more than adequate quality for high-resolution television production. In addition, zooms are much simpler and handier to work with—especially when the DP does not have one or more camera assistants to quickly retrieve the proper lens and mount it for her.

For the director, each of these lenses with their varying focal lengths means different types of frame sizes and thus different visual qualities for his shots.

Shot descriptions The frame sizes used most commonly in corporate television describe the following types of shots: *wide shot (WS), medium shot (MS), medium close-up (MCU), close-up (CU),* and *extreme close-up (ECU).* Variations of these frame sizes also make up other commonly used shots, such as "two shots," over-the-shoulder shots, point-of-view (POV) shots, and reverse-angle shots. Since a director must be knowledgeable about these shots, let's briefly discuss each.

Wide shot (WS) A wide shot is normally made with a mid-range focal length—often around 30 millimeters. It usually encompasses all of the action within a scene and establishes the area or location in which the action is taking place. A wide shot normally also keeps everything in the frame in focus.

A wide shot might allow us to see a complete den or kitchen, a full set, or a work area in which employees or actors are working. In our previously diagrammed shot of the service representative and her supervisor, camera position 2 would produce the wide shot shown in Figure 7.2. It would serve the purpose just mentioned.

Medium shot (MS) A medium shot most often frames one actor from about the thighs up. A medium shot can be framed nicely with a 50-millimeter focal length. Its purpose is often to feature an actor but not call dramatic attention to her.

A medium shot is also used when a director wants to single out an actor but still keep the audience aware of their immediate surroundings or set elements. This is because, like a wide shot, it too often keeps the entire frame in focus (see Figure 7.3).

In our representative and supervisor scene, the over-the-shoulder angles, such as the one pictured again in Figure 7.4, are also medium shots. In this case, they might be called over-the-shoulder medium shots or just plain over-the-shoulder shots; either distinction is correct. In both cases, if the foreground actor were removed, the remaining actor would be featured in a medium focal length.

Medium close-up (MCU) Medium close-ups frame an actor from the lower chest up. A 75-millimeter focal length is often about right for a

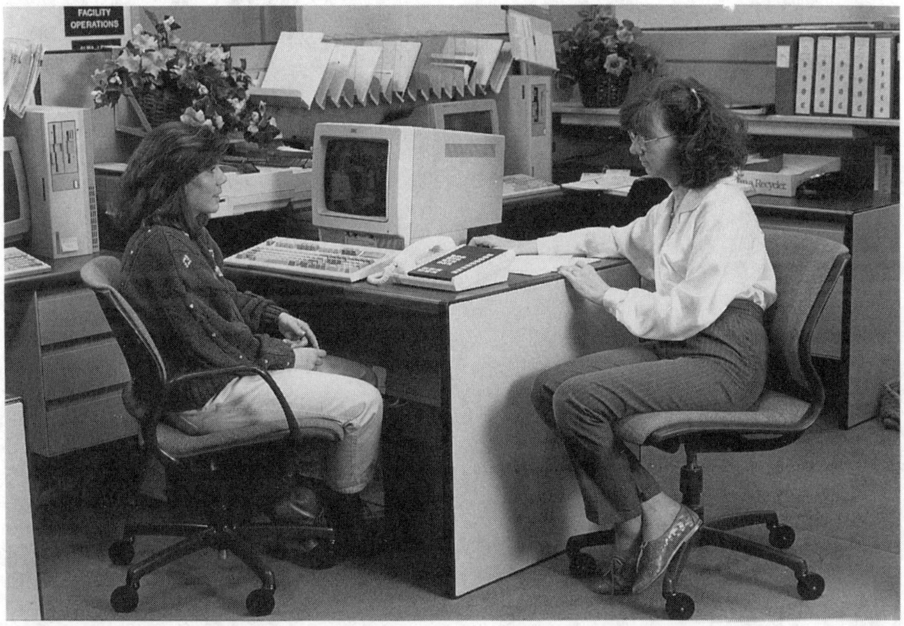

Figure 7.2 Wide shot establishes actors and surrounding work area.

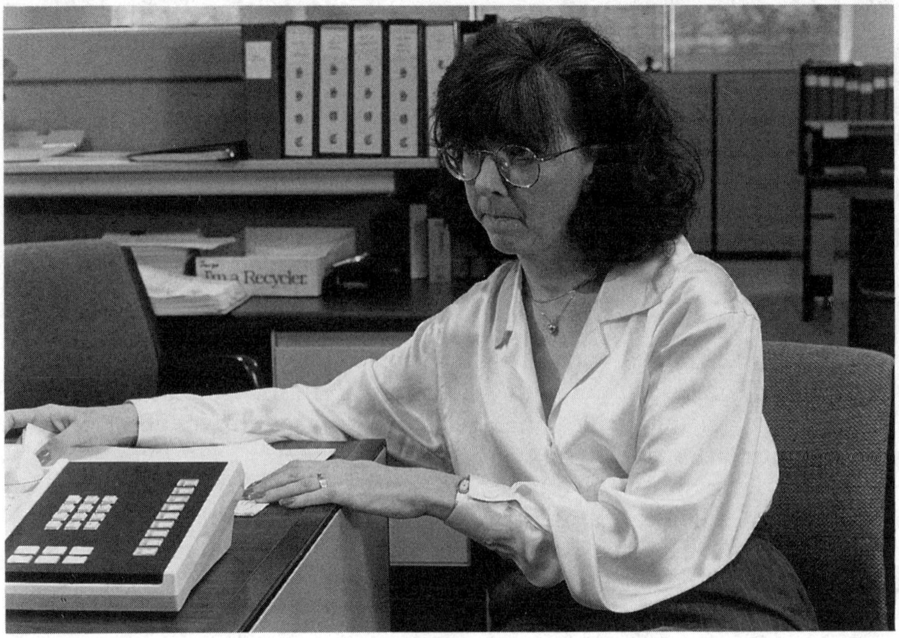

Figure 7.3 Medium shot singles out an individual but still gives a sense of the surrounding location.

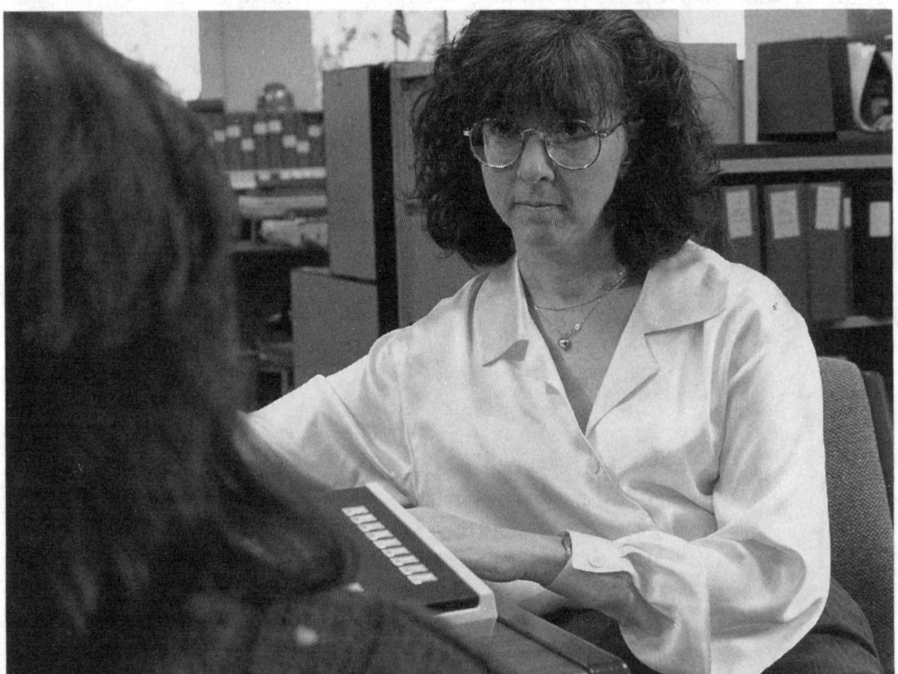

Figure 7.4 Over-the-shoulder shot (OTS) provides more depth and perspective than medium shot, since both actors are seen in the frame.

medium close-up. In this focal length, the background will often begin to go "soft" or lose focus, since the depth of field is more shallow than with a wide angle.

A medium close-up singles out the actor (Figure 7.5) and reduces the viewer's awareness of the surroundings. Medium close-ups are frequently used by directors to intercut individual dialogue segments between actors in conversations. Like medium shots, medium close-ups usually do not call dramatic attention to an actor, although they do bring audience focus strictly to her.

Close-up (CU) Close-ups should be reserved for dramatic emphasis. A close-up might be shot with approximately a 100-millimeter focal length. It fills the entire frame with the actor's face and perhaps upper shoulders (Figure 7.6). Close-ups typically also cause what little of the background is still seen to lose complete focus, so that the actor is separated still more from the surroundings. Most times, there is little motivation to bring the audience in to this intimate a perspective on an actor unless something she is saying must be punctuated or a subtle facial expression must be emphasized.

Extreme close-up (ECU) An extreme close-up fills the frame with the actor's face or part of it, as in Figure 7.7. It is normally shot at about 100 millimeters. Its purpose is to create dramatic effect or unquestionable emphasis.

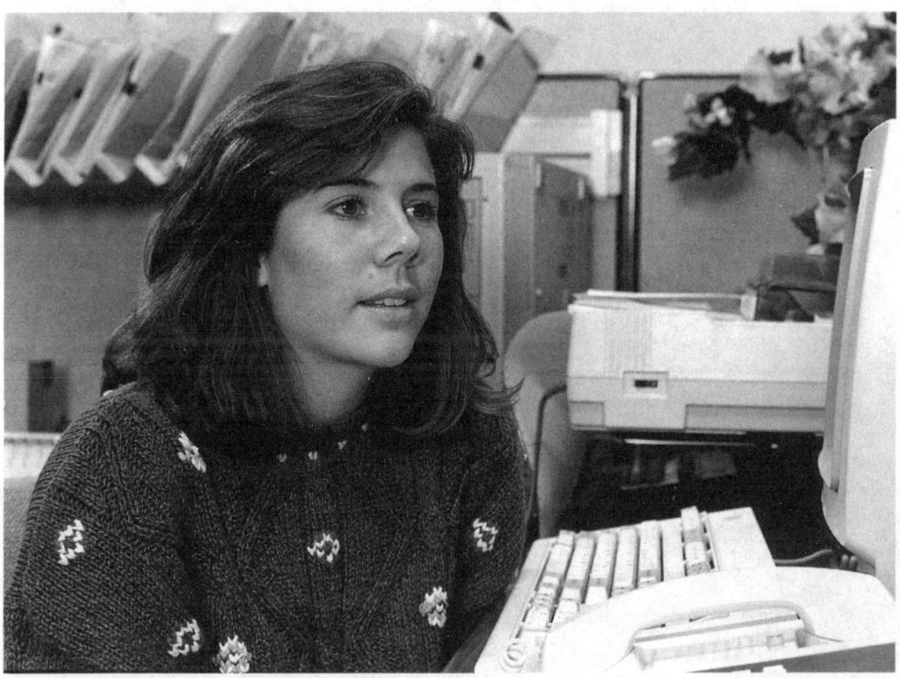

Figure 7.5 Medium close-up features an individual and begins to lose background.

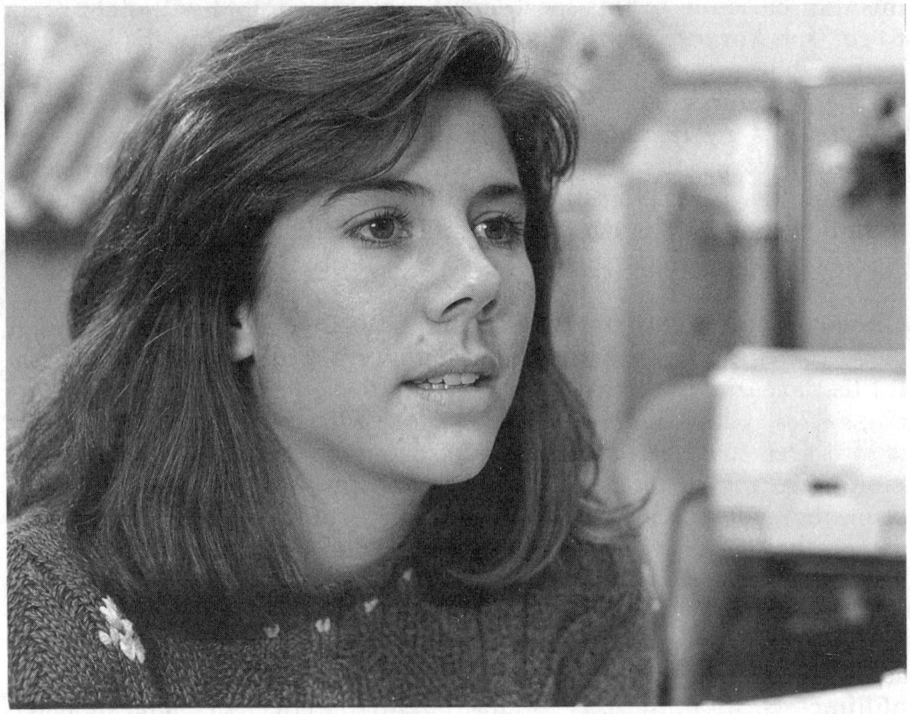

Figure 7.6 Close-up commands full attention.

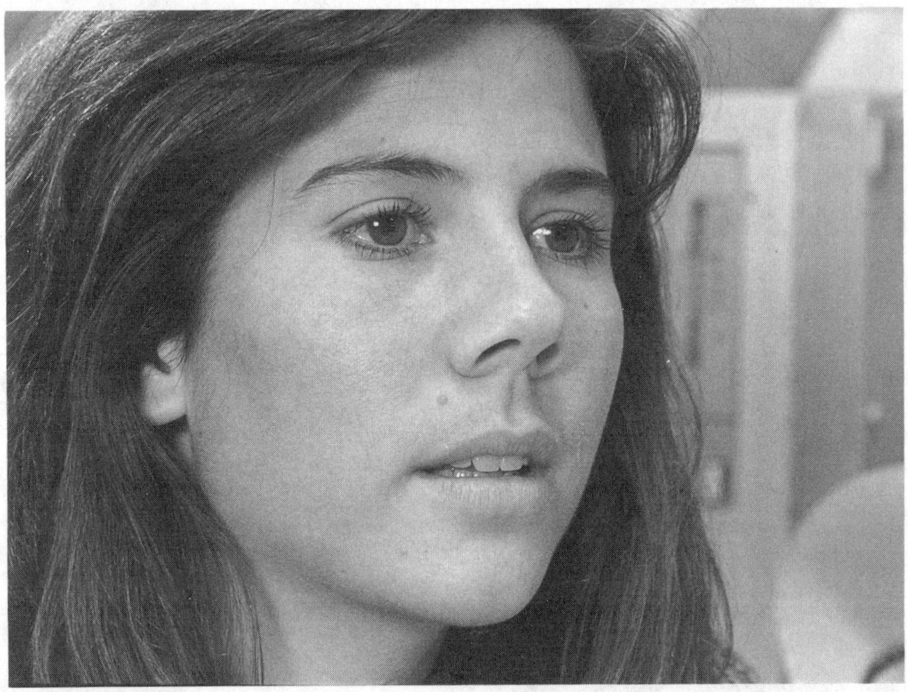

Figure 7.7 Extreme close-up is often used for dramatic emphasis.

An extreme close-up can be used to see fear, anger, or perhaps tears welling up in an actor's eyes. It can also call attention to a lip quiver, a twitch, or a toothpick being rolled over a tongue.

As with close-ups, extreme close-ups should be reserved for those few instances when the audience will not become aware or be made uncomfortable by being placed, in effect, nose to nose with the actor.

Extreme close-ups of objects or special actions are often called *inserts.* Inserts are typically shot to highlight some small or important action or element of the scene.

For instance, in our representative and supervisor scene, when the representative presses and releases the hold button (pictured again in Figure 7.8), the coverage includes an extreme close-up or insert showing this action take place. If part of this program's purpose were to instruct employees on when and when not to put customers on hold, use of these insert shots might be appropriate to ensure that the audience had noticed whether the representative followed the procedure correctly or not.

By contrast, if the scene had nothing to do with putting customers on hold and the action were simply incidental, the insert would be unnecessary.

Avoiding "Jump Cuts"

Another frame size consideration the director must keep in mind when shooting coverage of any scene is to avoid what are referred to as "jump cuts."

A jump cut is an unsightly edit that occurs when two similar pieces

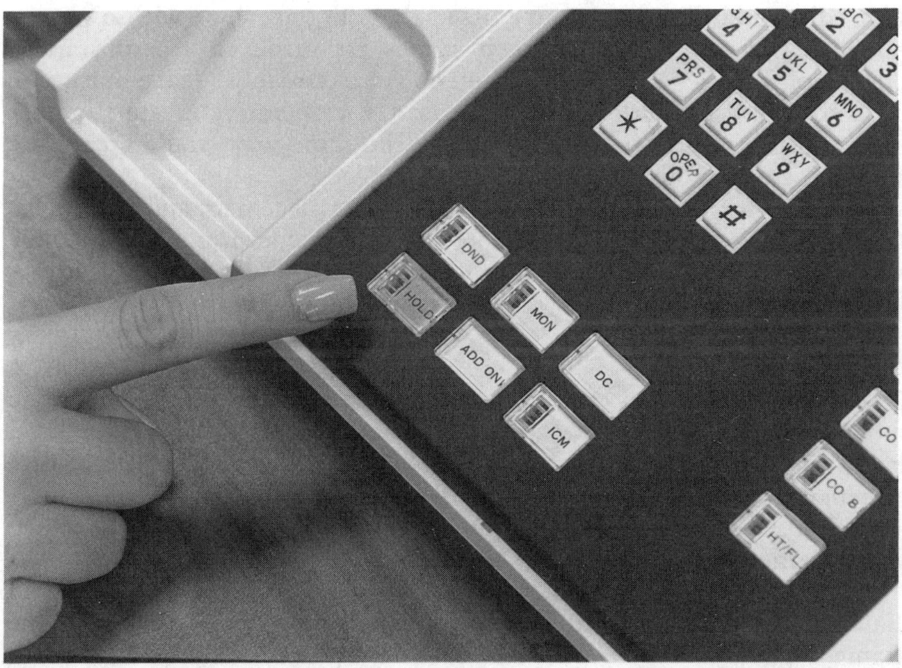

Figure 7.8 Extreme close-up of an object such as telephone is often labeled an insert. This is the case when it is "inserted" into the main action of the master shot.

of footage shot from basically the same focal lengths and usually the same camera positions are cut together. These types of junctures create pictures in which the elements in the frame such as people or objects appear to jump suddenly from one position to another.

Jump cuts can be avoided by keeping the following basic parameters in mind when shooting more than one angle of any scene or shot.

Change focal lengths Never shoot coverage of the same action without at least a 15-millimeter change in focal lengths. For instance, a host faces the camera in a medium close-up and records scene one. The camera is stopped when the director says cut. The camera rolls again for scene two and the host is still framed in a medium close-up. The shot is recorded. When these two scenes are cut together, the host will appear to suddenly jump from one position in the frame to another.

Change camera positions Try to change camera positions by at least 40 degrees when shooting coverage of the same action—even in addition to change of focal length. Using the same host example just quoted, scene one is shot in a medium close-up and the camera is stopped. The focal length is changed to a wide shot, but the camera remains in the same position and scene two is recorded. The footage will cut together acceptably, but the direct jump from the narrow to wide focal length might make for a somewhat uncomfortable transition. On the other hand, if the host had turned to a new angle between scenes one and two the camera would record not only the change of focal length, but also a changed background, which (provided the continuity of the turning action matched) would result in a very comfortable transition.

Plan edits carefully Plan your edits so that not only shots within scenes, but also transition edits between scenes are made with a substantial change of focal length or camera angle. For example, to leave one scene on a shot of a person looking just past the camera and begin the next scene with a person framed and facing the same way would create the same type of uncomfortable visual jump.

A director who has a good sense of camera placement and movement and who then becomes comfortable with the proper frame sizes for his subject must next address what will appear in those frames and how it will be arranged—in other words, composition.

Frame Composition

To compose means to create. Frame composition is exactly that. The director, working closely with the DP, arranges actors and foreground and background objects in the frame to *create* an overall visual effect that enhances or provides credibility to all other aspects of his illusion.

Frame composition is a subtle skill that usually involves three primary elements: *balance, depth,* and *mood.*

Balance Balance refers to how people and objects are arranged in the frame. Figures 7.9, 7.10, and 7.11 show examples of balanced and unbalanced frames.

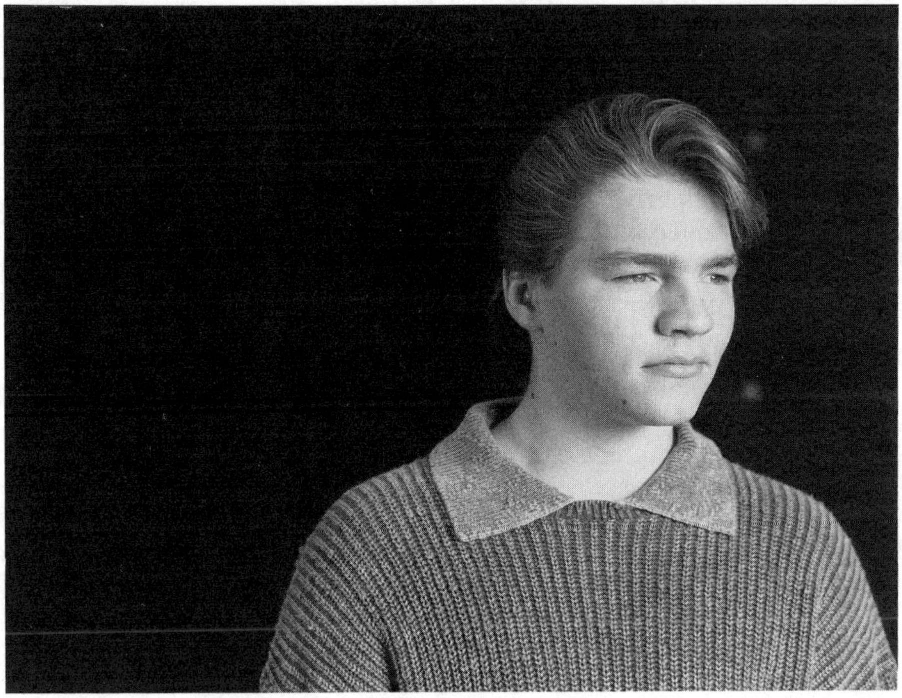

Figure 7.9 Unbalanced image. The face is up against the leading frame edge, and distracting space is left behind the actor.

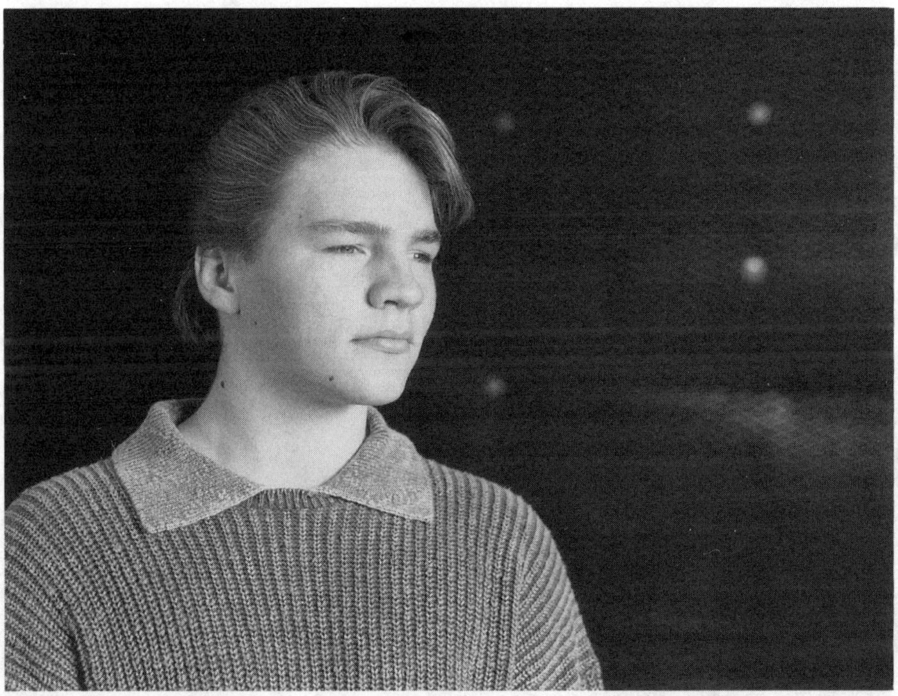

Figure 7.10 Balanced frame. The face has lead space in the direction of attention and is comfortably placed slightly to the left of the overall frame.

In Figure 7.9, the actor's face and upper body are at the right side of the frame. This leaves a good deal of empty space at frame left. In addition, his eye line is directed toward the edge of the frame he is *closest to.* This creates an uncomfortable or unbalanced feel that is distracting to the viewer.

In Figure 7.10, however, the frame is comfortable to the eye. The actor is framed slightly to the left of center, facing screen right. He appears naturally positioned in the frame, and his eye line is given a bit of "lead" space, rather than being "up against" the left frame. The viewer will have no trouble at all watching this picture.

Understanding this type of balance gives us some sense of composing the left and right sides of the frame, but what about the top and bottom? These areas too help make up the sense of visual balance.

In Figure 7.11, the frame is unbalanced because there is far too much headroom in the picture. Comfortable headroom can normally be achieved by keeping the actor's eyes roughly one-third of the way down in the frame. This will usually leave a small but very visually comfortable space between the actor's head and the top of the frame. In extreme close-ups, the top of the actor's head is often cut off. This is called giving the actor a "Warner Brothers haircut."

This one-third rule, like most others in the art of directing, is not without exception. For instance, an actor might be shown in relation to an area or object that requires him to be at the bottom or perhaps the top of the screen. This is often the case when logos, buildings, or other elements must also be framed in the shot.

Shots of places and objects should also follow the rules of balance.

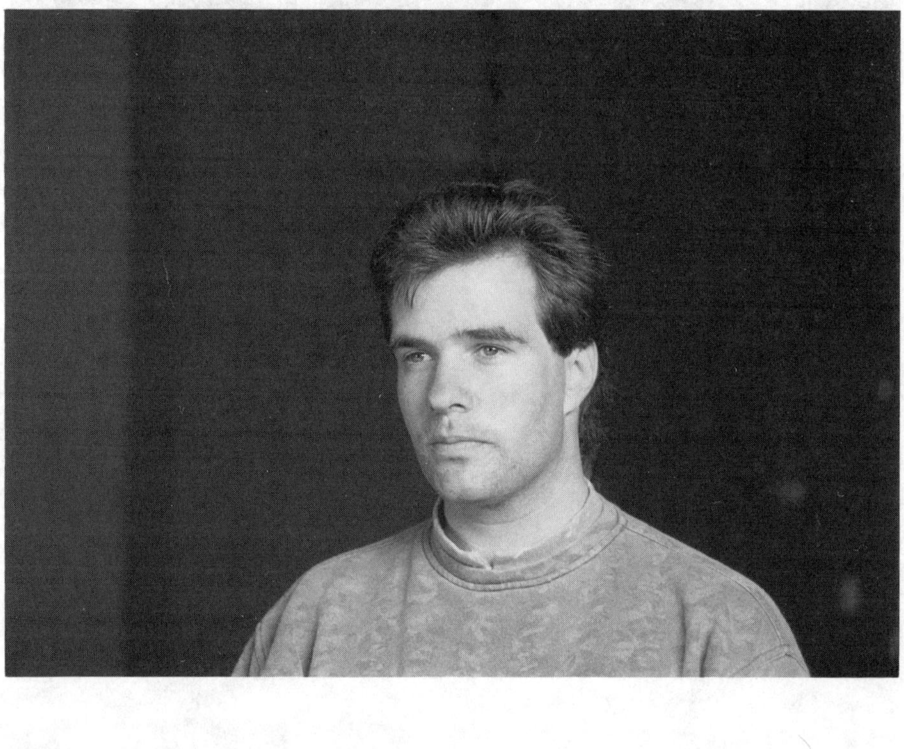

Figure 7.11 Too much headroom creates a very displeasing and uncomfortable composition. Eyes should be approximately one-third of the way down in the frame.

For instance, a horizon should almost never be made to fall in the middle of a picture, as in Figure 7.12. This tends to flatten the image and destroy the viewer's sense of perspective and depth.

In Figure 7.13, this is corrected by placing the horizon in the upper third of the frame, thus making the foreground appear much more interesting and comfortable to the eye.

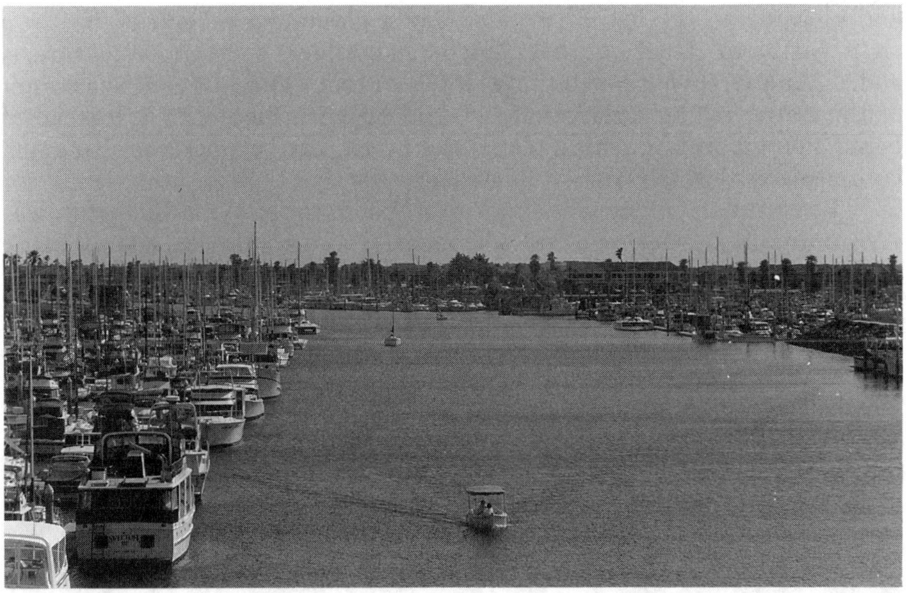

Figure 7.12 Horizon in middle of frame creates a flat and uninteresting image with no emphasis on foreground or background.

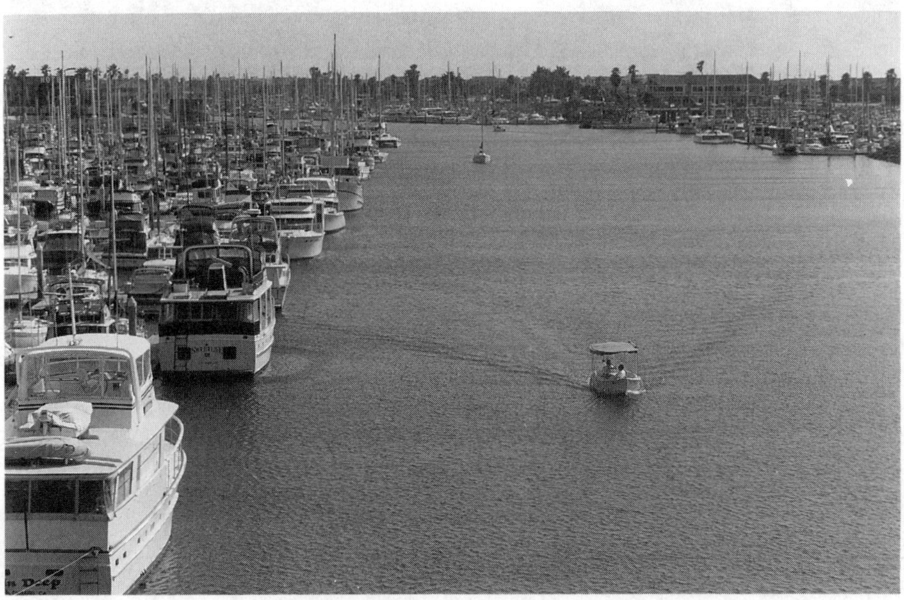

Figure 7.13 Horizon at lower or, in this case, upper third area of image is much more pleasing to the eye. Foreground becomes area of emphasis.

Depth With a basic feel for the sides and top and bottom of the frame, a director must consider *depth*—the foreground and background. Foreground or background elements, or the lack of them, can make a picture seem flat and uninteresting or provide an aesthetically pleasant feeling of depth and texture.

For instance, in Figure 7.14, an employee being interviewed in front of a flat, pale wall produces an uncomfortable and even somewhat unnatural composition. If a small tree is placed in the *background* over her right shoulder (screen left), the picture becomes a bit more interesting and pleasing to the eye. If we also add a globe and several books on a shelf partly in the frame over her left shoulder, even more texture is added. Finally, if we position several out-of-focus fern leaves at the corner of the frame in the *foreground,* we end up with Figure 7.15. Instead of being a single uninteresting plane, the image has become a multiplaned composition that is visually balanced to the eye.

By contrast, you may be shooting a host in limbo (a totally featureless environment) presenting a very serious subject. In this case, a stark black or white background might be appropriate to the content being delivered.

Depth can also be created simply by not shooting against flat surfaces. In an office or set, it is often visually more interesting to shoot so that corner can be seen in the frame or so that wall areas are framed diagonally. Again, this adds a pleasing sense of perspective and depth.

It is this sense of depth that often motivates directors to shoot over-the-shoulder and two shots instead of individual medium close-ups of

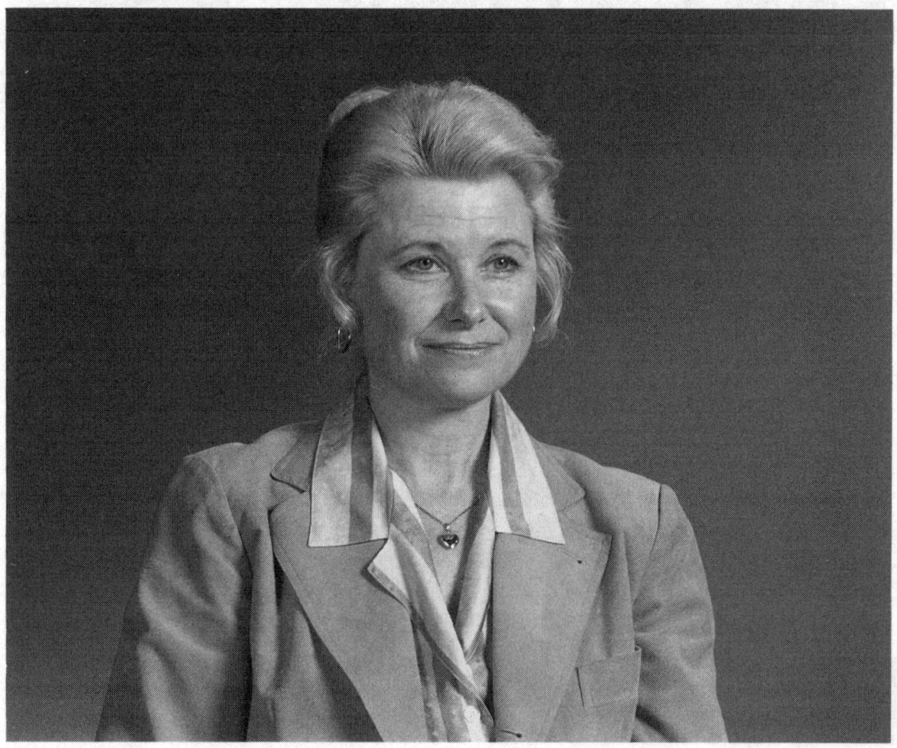

Figure 7.14 Employee interview composed with flat, uninteresting background.

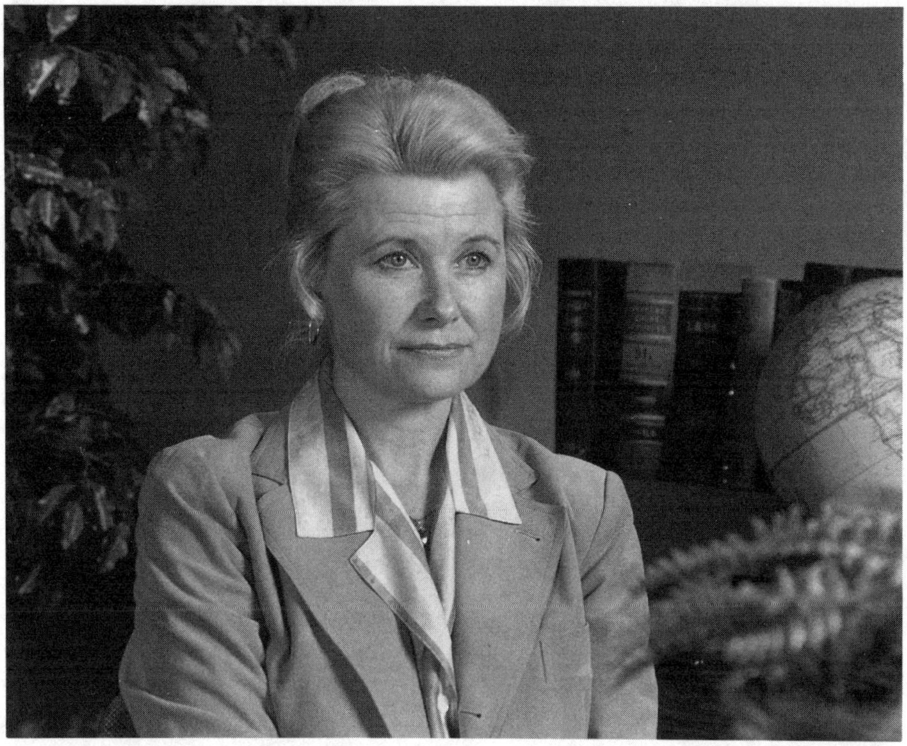

Figure 7.15 When plant at left is added, pleasing shapes help provide visual texture. Adding books and globe provides still more depth and interest. Leaves in foreground add another level of depth, completing the "composition."

actors. While a medium close-up may focus in on the actor, as in Figure 7.16, an over-the-shoulder shot can also establish a visual sense of relationship between the two actors—in other words, a feeling of depth (Figure 7.17).

Depth can also be controlled by lens usage and light levels. Lenses with long focal length, as we have said, tend to flatten a background and foreground, making distant objects appear much closer to each other than they are. Wide-angle lenses do the opposite.

F-stop also affects depth by changing the depth of field. When a camera's aperture or iris is fully opened, in low light conditions, the depth of field is reduced sharply. This means that foreground and background objects lose focus if they are not close to the actor. A nearly closed F-stop, on the other hand, used in bright lighting conditions, increases the depth of field so that background and foreground objects appear in focus, even if they are quite a distance from the actor.

Mood Mood is created by a number of picture and sound elements, one of which is frame composition. Consider the two pictures in Figure 7.18. Both represent essentially the same content—an employee interview. The first one (A), however, creates a bare or sterile mood with its simple vertical lines. The second picture (B), by contrast, creates a more positive, interesting mood, simply because it has shadow clusters that break up the vertical lines, creating more depth and subtle interest.

As you can see, then, what is placed in the frame and how it is

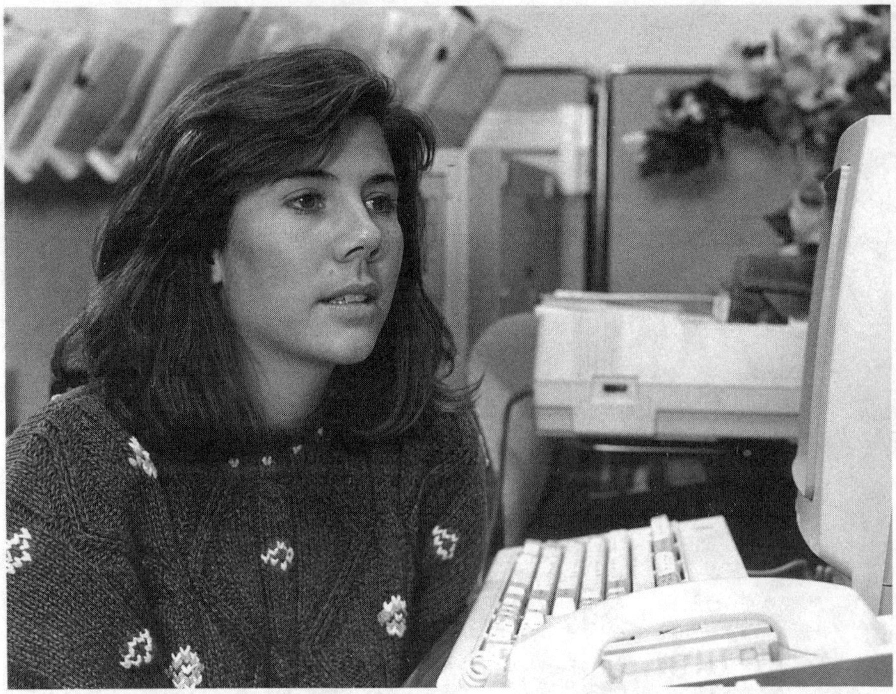

Figure 7.16 Medium close-up.

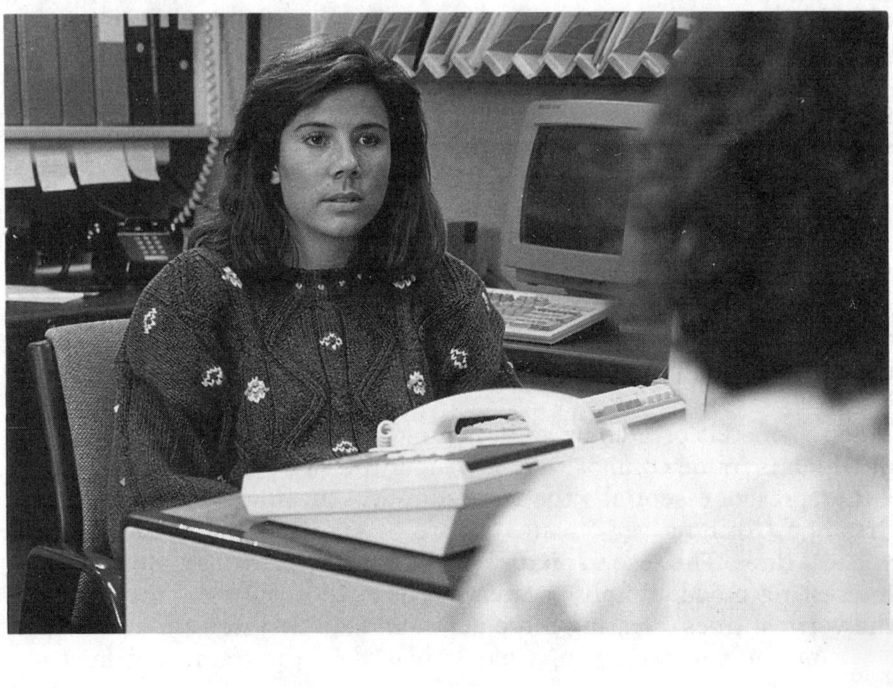

Figure 7.17 Over-the-shoulder shot.

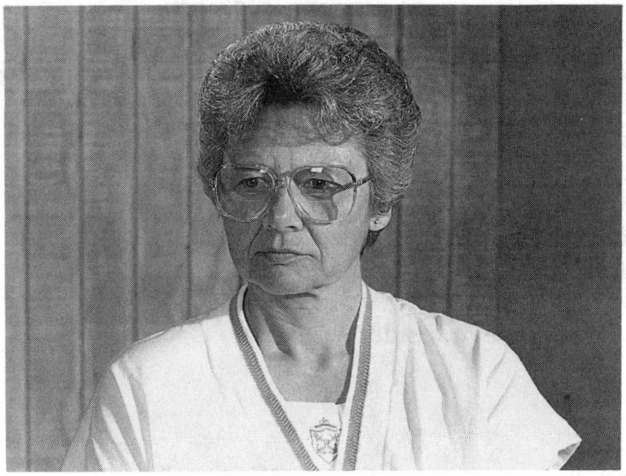

A

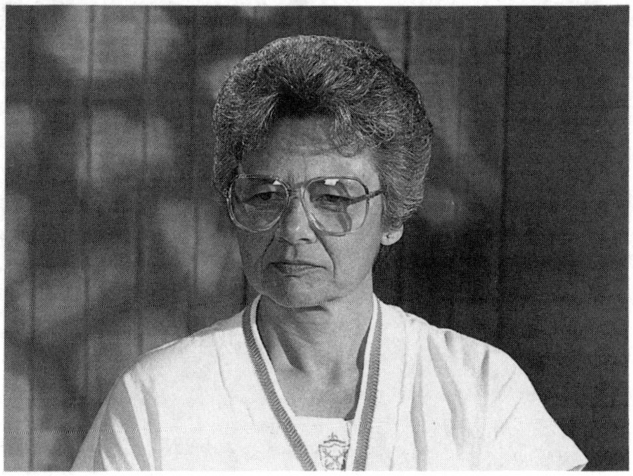

B

Figure 7.18 Straight vertical lines provide a sterile look (A). Shadow clusters "break up" the vertical and provide visual interest (B).

arranged have enormous subtleties and implications in the pictures a director shoots. In these examples we have just scratched the surface of this aspect of the director's work. Still another element of audiovisual aesthetics that has a profound impact on the visual elements of the program is its lighting.

Lighting

Lighting serves two important purposes for the corporate video director. The first is simply to make the scene being shot bright enough to achieve the proper video levels. This ensures that the technical standards are met and that the visuals are of adequate quality to be reproduced, edited, and, of course, viewed by an audience. The second purpose of lighting is to enhance the images being shot and create moods and tones that support the director's illusion.

Achieving the proper video levels and values As mentioned earlier in the section on production lighting, the brightness of a video scene is measured in terms of luminance. Luminance is measured on a waveform monitor that is either taken to the field or mounted in a studio control room.

The brightest parts of a video picture—usually the whites—should register at approximately 100 units on the waveform monitor. The darkest parts of the picture—the blacks—should register at 7.5 units.

In addition, the overall video image being shot should not contain severe contrast between these two extremes. Instead, a good video image will normally contain a *range* of levels in between 7.5 and 100 units. A human face, for instance, will normally register between 60 and 70 units. A dark shirt or suit will register perhaps 30 units. A pale wall may register 90 units.

If a part of the picture becomes overexposed or registers over 100 units, the video signal will become distorted and that portion of the image will seem to "bloom" unnaturally. These overexposed portions of a picture are usually called "hot spots."

If too much of the picture is extremely dark, it may appear acceptable on the high-quality broadcast system that is recording it, but later, when it is duplicated to a viewing format such as standard VHS, the darks will begin to break down, creating grainy or "noisy" images, as they are often called.

Achieving these technical standard levels while creating moods and textures with lighting can sometimes be difficult. It is important that the director understand these issues because they directly affect how he might plan to shoot a scene. For instance, consider this scene:

INT CEO'S OFFICE—DAY
The CEO is standing at his large window, gazing out over the city, as he carries on a conversation with one of his VPs.

CEO
John, I'm convinced we should up the stakes on the Waite project. They're going to blow this thing sky-high unless we get in on it now.

Remembering what we just said about not having too much contrast in a video image, would it be a good idea to plan on shooting this scene with the window as a backdrop for the CEO? It's doubtful. The outside light would probably be much too "hot," surrounding the most likely dark-suited figure of the CEO in front of it.

If forced to shoot such a scene, the DP would probably start by closing down the f-stop of the camera until the outside light registered 100 units on the waveform and did not "bloom." At this point, the background luminance level would be proper. But with the camera iris shut down this far, the dark-suited image of the CEO would probably appear only as a silhouette, since this portion of the picture would be severely underexposed.

To balance the overall image, the DP would have to focus a good deal of light on the CEO so as to "pump him up" to match the background level. Although this would probably produce the proper recording levels, there's a good chance it would also make the CEO look very lit and unnatural.

A better solution would be for the director to choose a camera position that does not juxtapose the CEO in front of the window. Shooting him from the side, for instance, with a small portion of the window in the frame and a hard key light that appears to be outside light, would no doubt create an effective look. It would also save a good deal of time and result in a much more natural and aesthetically pleasing shot.

The same types of problems can develop when dealing with a *lack* rather than an overabundance of light. It is for these reasons that the director should be well aware of the capabilities and restrictions of his lighting needs.

Creating mood, tone, and textures Lighting also helps create moods that strengthen the director's illusion. Here's one simple example:

INT SET—DAY
We see only a dark wall. On it is a shadow pattern that appears to be light shining through bars in an open jail cell door. HOST ENTERS under what appears to be harsh, single light source.

HOST
(Gestures to shadows)
Life under lockup. Not a very settling thought. But is it really that bad? This is "Newsline," and in this issue we're gonna show you.

As host exits, the shadows on the wall begin to move, suggesting that the cell door is sliding closed. It slams shut with a METAL CLANK.

Obviously, the dark walls and cell door shadows help create the tense mood the director wants for the "open" of this program. The "single" light source on the host also helps reinforce the dark, harshly lit conditions that might exist in a real jail cell. Both elements, in concert with a convincing performance by the host, would create a very credible opening segment that would no doubt keep viewers waiting for more.

Standard 3-point lighting In many corporate programs, a dramatic mood is not needed. What *is* needed, however, is an attractive image, complementary for the actors or employees being shot. This might be the case when shooting a videotaped message by a company president who must look very corporate and polished. Or it might make an employee interview—say, in a supply room or manhole—appear natural but aesthetically pleasing for the location. When this is the case, variations of a standard 3-point lighting setup become the tool used by the DP.

Three-point lighting refers to a "key" or primary source of light, a "fill" or weaker source used to fill in shadows created by the key light, and a "back" or "rim" light used to separate the subject from the background by creating a thin rim of light on the subject's head and shoulders. Figures 7.19 through 7.23 show a number of variations on standard 3-point lighting setups and diagrams explain how the different types of

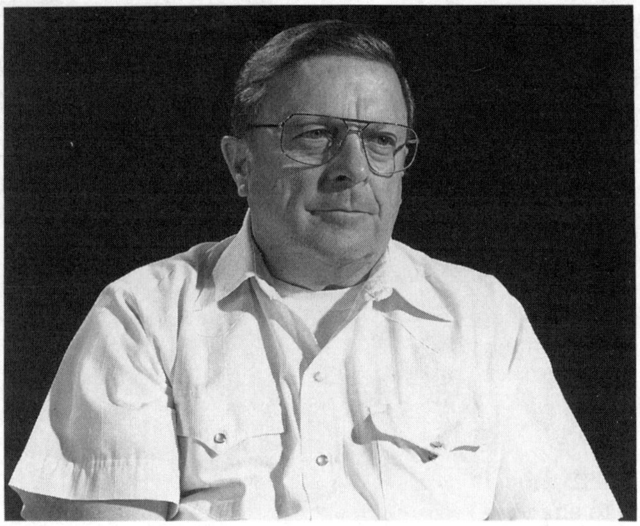

A

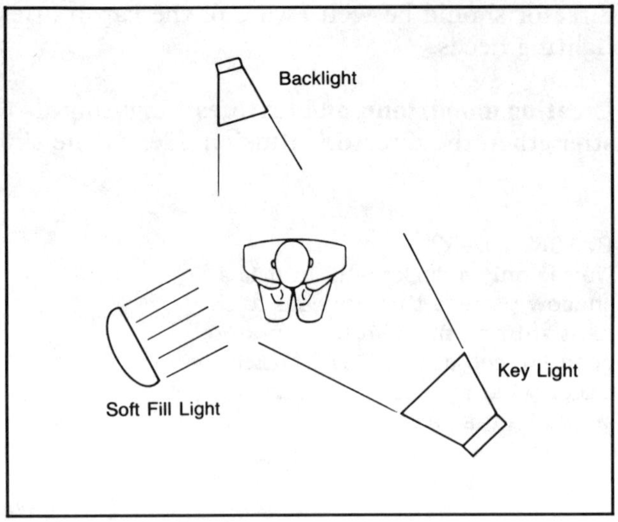

B

Figure 7.19 Interviewee lit with a standard 3-point lighting setup—a hard key light, a soft fill light, and a backlight.

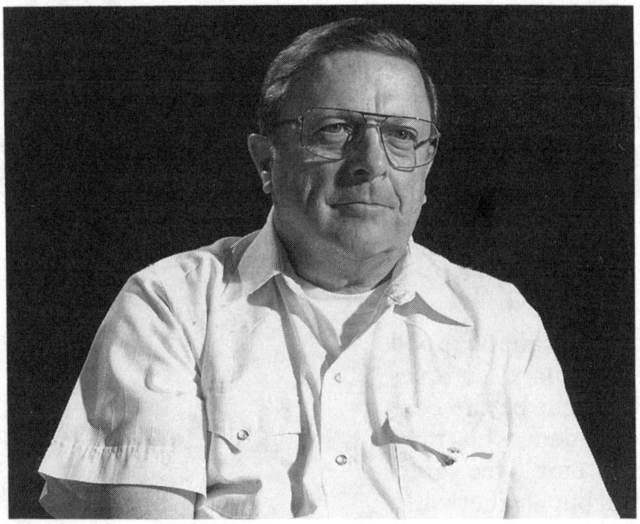

A

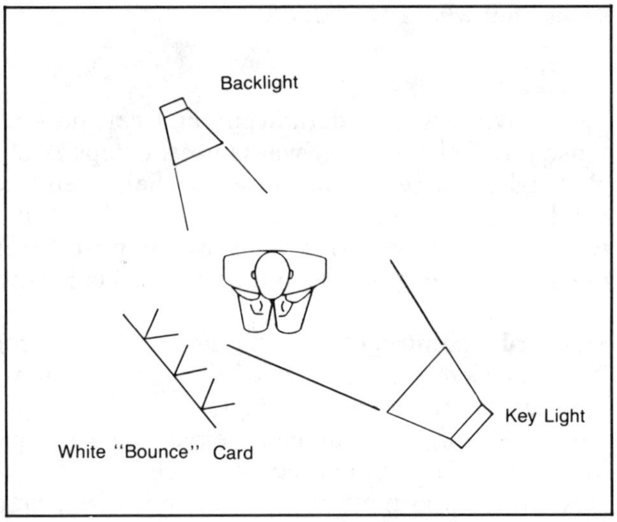

B

Figure 7.20 Same setup, except fill light has been replaced with a bounce card, producing more shadow.

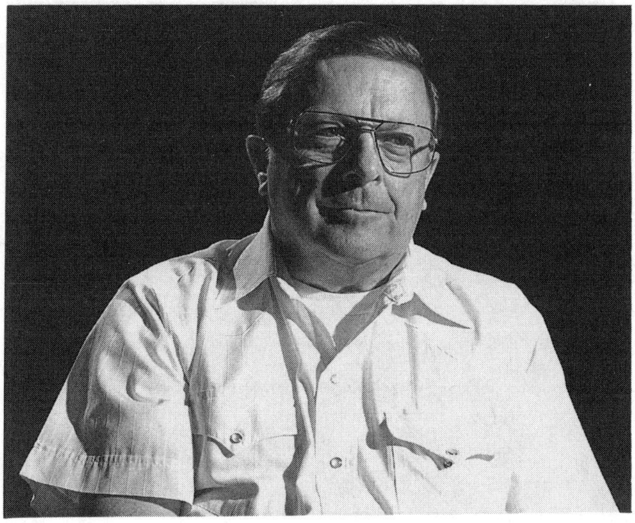

A

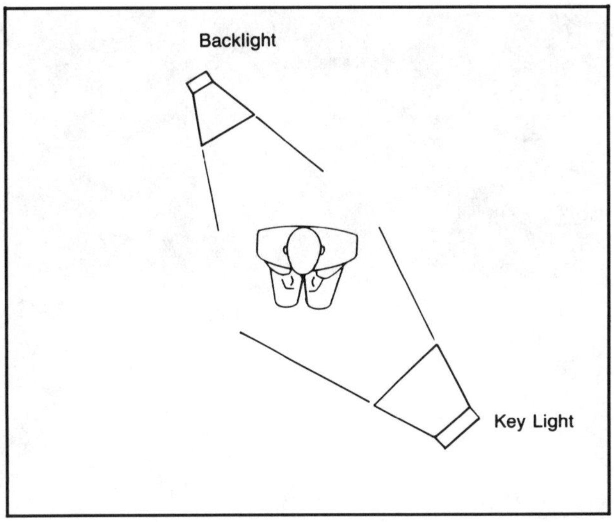

B

Figure 7.21 Same setup, but in this case with no fill light. Shadows are now becoming dramatic.

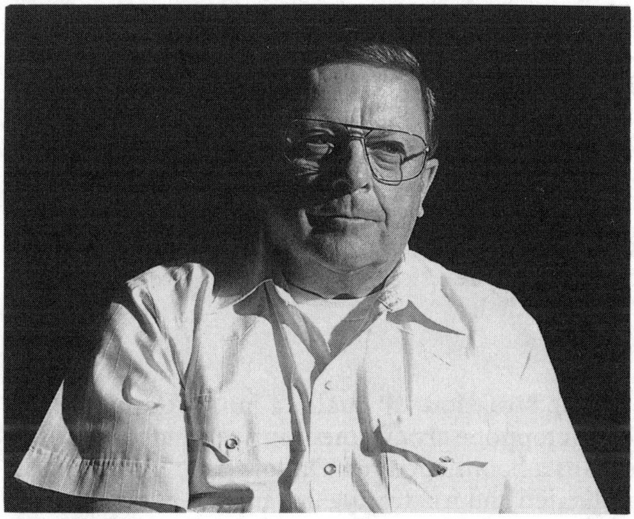

A

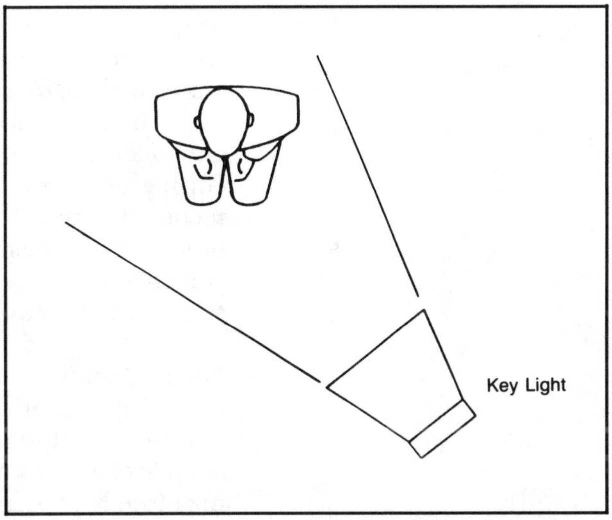

B

Figure 7.22 Now only a single hard source of light is being used—the key light.

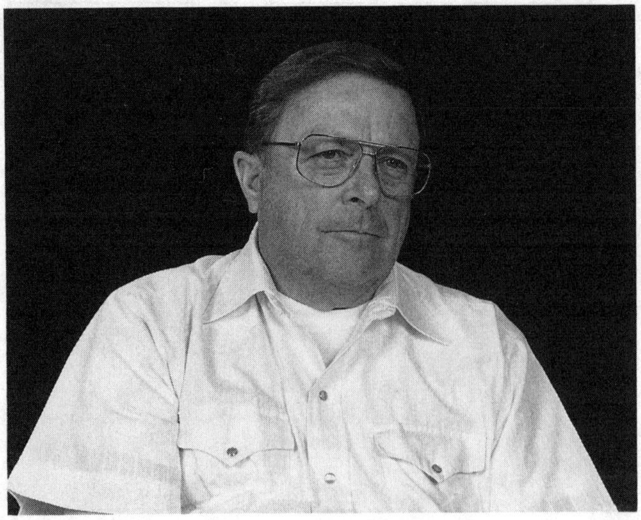

A

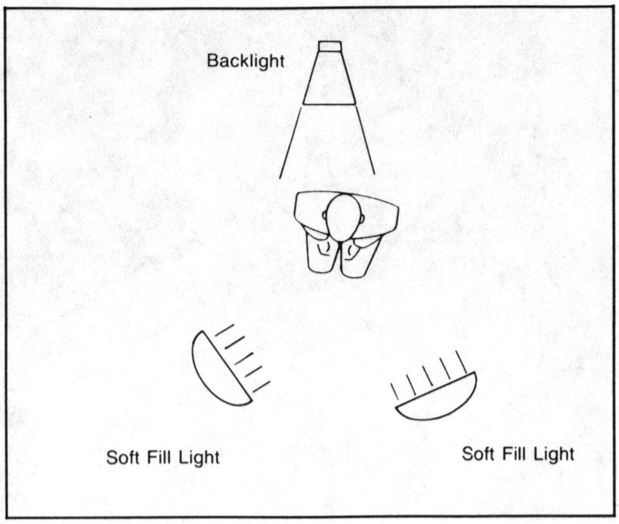

B

Figure 7.23 Flat lighting (no shadows) is created by using a wash of soft equal light.

looks are produced. Again, the director must be aware of these lighting options as he considers how he will shoot the various segments of his program.

Sound

The audio or sound portion of a production also has a profound effect on its aesthetic quality and audience impact. The right sounds, creatively placed and appropriately enhanced, can wake up an otherwise "flat" scene.

Unlike lighting and camera work, however, the majority of sound enhancements on a typical production happen after production, in the editing or audio "sweetening" rooms. These processes enhance existing recorded sounds and add sound effects and music at the director's or producer's request. During the production process, there are two primary considerations for the director in relation to sound: the *quality* and the *types* of the sound being recorded.

Sound Quality

Recording sound in a studio situation is usually a fairly routine matter that involves the proper microphone arrangement in the hands of a skilled sound recordist and perhaps a boom operator. On location, sound recording often becomes complicated and frustrating due to elements out of the production crew's control.

As an example, I once shot several host scenes in a house in a residential area. In one, the host was to step into a kitchen and introduce the subject of home security systems.

Due to a camera problem, we had started out behind schedule that day. Next, a conference about a wardrobe change between me, the actor, the producer, and the client held us up even more. On top of all this, the lighting was fairly time-consuming because we were shooting day for night (a scene shot during the day that has to appear as if it were happening at night).

When we had at last managed to iron out these problems and rehearse the scene a few times, we were over an hour behind schedule. I finally decided we were ready and said, "Let's role tape and try one."

The actor prepared. The crew settled. The slate was held in front of the camera. Just as the camera rolled, however, the whine of a distant lawn mower suddenly started up. I looked at the audio recordist, who had on a set of earphones to monitor the sound being recorded. She returned my glance and shook her head. This meant there was no way we could record the scene. The whine was prominent enough to be recognized and also to affect the sound continuity of any editing that might take place later.

What did we do? The AD located the gentleman mowing his lawn and offered him $20 to put off his yard work until the afternoon. Luckily, he accepted.

This is just one example of numerous types of sound problems that can and often do arise when shooting on location. Whether in the studio or on location, a director should keep in mind a few common rules of thumb and concerns.

Microphone choices First, shotgun mics mounted on "fishpoles" generally produce high-quality audio, but they typically require a two-person sound crew—a mixer and a boom operator. The boom operator follows the actors around, holding the fishpole, so that the mic is "on axis" with the actor speaking. The mixer or recordist monitors the sound at a mixer and makes the necessary adjustments to the incoming levels to be sure the recording is technically acceptable before reaching the VTR.

In some cases, the boom operator is not needed if the actor will remain in a single spot. In this case, the fishpole can be mounted on a "C" stand or light stand and locked into position just out of the camera's frame.

Another possibility, if the actors are to remain fairly immobile, is to use lavaliere mics on hardwire cords or RF transmitters. As we have seen, lavalieres are extremely small capsule mics, and they can be easily hidden under a collar or inside a shirt. They too produce high-quality audio but typically begin to make rustling noises when actors move about and their clothes begin to shift. This is especially true when the actors are wearing clothes made of silky materials, such as ties and thin blouses.

A hardwired lavaliere clipped to a tie or lapel and used in an "exposed" fashion is often excellent for employee interviews and host scenes when a wide angle will *not* reveal the attached audio cables to the audience.

When audio cables may end up being seen in a shot, lavalieres on RF transmitters usually solve the problem by eliminating the cable.

Types of Sounds

A director should always plan to record ample, good-quality dialogue and narration, as well as *ambient* sounds.

Ambience To the TV or film sound recordist, ambience means sounds present or characteristic of the location. In a gym, ambient sounds would be weights clinking, the chatter of voices, and perhaps the sounds of people straining or breathing heavily. On a city street corner, ambience would be traffic sounds, horns, and perhaps pedestrian chatter. Even rooms have ambient sound characteristics—often called "room tone" or "room ambience." A few minutes of ambience should be recorded at every location the production crew visits.

These ambient sounds should be recorded *separately* as should dialogue and narration so that *all* recordings are kept totally isolated or "in the clear." This allows them to be edited together later with maximum effect and flexibility.

The aesthetic qualities that well-recorded sounds bring to a production can't be overemphasized. They add a subtle level of enrichment and an underlying texture to a scene that, just like all other audiovisual aesthetic qualities, provide still more credibility to the director's illusion.

Summary

If we take a moment to consider what we've covered up to this point, we should recall that

- First the director analyzed his script and from this analysis made numerous, initial decisions about actors, locations, camera positions, lighting, and so on.
- Next he chose his actors and guided his cast in rehearsals toward the performances he had visualized.
- Through the use of audiovisual aesthetic qualities and with the support of his crew—in particular, the DP and sound recordist—he then ensured that the proper visual and sound elements were utilized to give the scenes maximum credibility.

It would seem that this is enough creative responsibility for any person. For the director, however, there are quite a few technical and general aesthetic qualities yet to be considered.

8

Technical Aesthetics

We have examined how important human and audiovisual aesthetics are to the corporate director. Just for the moment, let's assume she has done an excellent job in these two areas. The performances in her program are perfectly credible, as are all camera, lighting, and sound elements. But suppose some other, *seemingly* minor details are not so perfect—say, a prop is wrong for a job situation or a piece of wardrobe doesn't fit the setting or location.

Assume, for example, that an employee was supposed to have been using a specific, company-approved drill when he was taped. As it turns out, however, he was actually recorded using a different drill—one he had brought from home. Or instead of the company-mandated, steel-toed, leather shoes, he wore tennis shoes. Or—even worse—suppose he was wearing leather shoes in the opening scenes, but later, because certain scenes were shot on a different day, he wore loafers.

What effect could these kinds of errors have on the completed program? They could easily negate all the other positive things the director had done and ruin the entire production. That's because even minor *technical inaccuracies* or *continuity problems* can break the director's illusion and result in an instantaneous lack of program credibility.

Avoiding Technical Inaccuracies

Avoiding technical inaccuracies most often begins by paying great attention to a number of important details in preproduction.

Props

Most times, the job of lining up props for a corporate production is left to the AD. It is the director's job, however, to approve the acquisition of these elements, oversee the acquisition process, and make key decisions as to how, when, and where the props will be used. This process begins in the very first phases of preproduction and often continues through the production days themselves.

For instance, consider this scene:

```
EXT ALLEY—DAY
The installer dons his climbers
and work gear and starts up
the pole.
```

HOST (VO)
Having no luck in locating his line trouble in the house wiring, the installer may try the pole terminal . . .

The "work gear" referred to in this scene might include special kinds of tools. Neither the director nor the AD, however, would be made aware of these special needs by reading the script.

The wise director realizes this and makes careful and detailed notes during her script analysis. She then heads off such potential problems by asking the client or producer about any questionable items or areas. Or, if she is too busy taking care of other issues, she assigns the AD to research the questions and report back to her.

Based on this research, she then makes carefully thought-out decisions regarding the props required for each scene and their use. She also verifies that everything she has requested will be the correct, company-approved make, model, and issue and that it will be delivered to the proper location on the right day and time.

The corporate director who does not take full responsibility for these details often regrets her decision later. When she is sitting in the rough-cut meeting and the client brings up the fact that the type of screwdriver used by an employee or actor was not in accordance with company policies, the resulting decision may be to scrap the footage. This usually means a reshoot, which equates to additional time, money, and headaches for the producer.

Makeup

As previously mentioned, in feature film work actors are usually surrounded by staffs of people whose job it is to make them feel and look good. Making an actor look good is the job of a makeup artist.

In corporate television, makeup artists are often the exception rather than the rule. In many cases, thc actors themselves are asked to apply their own makeup, which usually consists of a simple cosmetic base and perhaps a thin coating of powder to smooth out wrinkles and hide blemishes or splotches.

When a makeup requirement does exist on a corporate shoot, it often becomes another of the AD's tasks. This is typically the case with executive shoots or employee interviews.

The director should be conscious of and prepared to handle makeup needs as one of her technical aesthetic responsibilities. This includes assigning the task to the AD if it is simple enough, or actually hiring a makeup artist if the situation calls for it. In either case, this is important for the same reason the director must approve props and wardrobe—because regardless of performance, camera work, or even very skillful lighting, a lack of makeup can sometimes ruin the credibility of a director's work.

To understand how this can occur, consider two inherent elements of the production process: lights, which generate a good deal of heat, and

Cheating: A Case of Limited Options

Case Study 5

I was once placed in a tense situation on a distant location shoot. I was to meet my crew and cast at a location some 30 miles from the studio, and they were to bring the client along for the day's recording.

When they arrived, only one of the two trucks we had loaded the night before showed up. After an hour's wait and finally a phone call back to the producer, we discovered that the second truck—containing a key prop we needed and several other important items—had mistakenly been hauled away by garage forces during the night. We could get it back, the producer said, but not until the next day.

The prop we were missing was a telephone terminal that was to be the focus of a discussion in several role-play scenes. The scenes involved actors paid for that day only.

Realizing it would do no good to panic or become upset, I decided that my best avenue was to take a short break and look over my options. I told the crew to relax with a cup of coffee, and I sat under a nearby shade tree, with my AD.

We carefully went through the script, the shot list, and our shooting schedule. After several minutes, the right course of action became clear.

We would shoot over-the-shoulder and close-up shots of the actors going through their scenes with a spray-painted, stand-in prop terminal kept mostly out of the camera's frame. We would also shoot establishing long shots of the interactions, conveniently placing what was supposed to be the terminal partly behind one actor. When viewing the establishing portion of a scene in a long focal length, I knew the audience would never realize it was being tricked.

Then, later, when the prop was retrieved, we would shoot the close-ups of what the audience would *think* were the actors' hands working on the terminal. The hands, however, would actually be those of stand-ins also.

We carried out this plan, and it worked perfectly. The next day, having retrieved the real prop, we were careful to match the action for our close-ups by first viewing playback of the previous day's footage. We also kept the frame fairly tight and carefully matched the lighting. We shot the close-ups in a grassy area beside the studio parking lot, and when the show was edited, no one ever knew the difference.

---

employees' nervousness when sitting in front of a camera. As you might guess, this combination of heat and nerves often leads to sweating—or getting "shiny," as the DP might say. When an employee or, heaven forbid, the CEO begins to sweat, that "shiny" forehead or upper lip quickly becomes a damaging distraction in the audience's eyes. The accomplished director avoids this and other makeup problems by being conscious of her makeup needs in preproduction and making absolutely sure those needs will be met when on location or in the studio.

Wardrobe

Employee wardrobe is often not a problem in corporate television *if* the employees are acting in their own job roles. An executive being interviewed or shot as an executive in a scene will most likely be wearing the appropriate dress. A customer representative or telephone installer will likewise usually be dressed appropriately for the job.

In these cases, the director must simply be sure the wardrobe meets all safety and company standards. The telephone installer, for instance, who shows up on the shoot wearing shoes or boots that are not company approved for climbing telephone poles will surely become a (possibly very expensive) problem later if he is not corrected on the spot. The trouble is that correcting him on the spot could result in an hour's downtime for the entire crew while the employee runs home to change his boots. Making the correction in preproduction, however, by having a brief telephone discussion with the employee or perhaps his supervisor, would head off this problem before it materializes.

When actors are involved in the shooting, it is the director's responsibility to decide on wardrobe. She does this in two ways. Initially, she usually speaks with the actors in preproduction and gives them a sense of what types of wardrobe will be required. This often happens in phone conversations between the two or by instructions given to the AD to pass along to the actor. In addition, the actor is expected to bring several choices of wardrobe on the first shoot or rehearsal day, and from these the director makes her final decisions.

In most corporate video productions, the rule of thumb regarding these decisions is *conservative* and *pastel.* In keeping with their physical appearances, actors' dress should typically be *un*noticeably natural for their surroundings and the situation. In addition, if they are to be shot in a location of predominant colors, the director may ask that they dress to complement the overall color scheme. Actors dressed differently from the corporate norm, especially in bright, highly saturated colors, often cause problems with lighting or video levels. If these technical issues do not result in trouble, the actor may simply become noticeably out of place and thus distracting to watch. This would not be the case if the program being shot called for outlandish or bright wardrobe as *part* of the director's illusion—perhaps in the case of a comedic approach or a festive party scene. As a general rule, the director should make it her business to be absolutely sure that all props, makeup, and wardrobe considerations are researched, confirmed, and scheduled in preproduction.

Continuity Problems

Another element that can have a distracting and often devastating effect on a program is a lack of continuity. There are two primary types of continuity a director must be concerned with: *physical continuity* and *performance continuity.* Both require maintaining close similarities between shots, scenes, and program segments. It is these similarities that trick the audience into believing that two shots recorded hours apart actually happened at the same instant.

Physical Continuity

To illustrate, let's briefly return to our scene in which the supervisor and a customer representative were having a discussion. Let's assume that throughout the shooting of the master scene, the supervisor was wearing a white blouse. Sometime after the master scene had been recorded, the crew and cast went to lunch, and it was not until the afternoon that the over-the-shoulder and medium close-ups of the supervisor were shot.

But suppose she had gotten a spot of mustard on her blouse at lunch and had changed to a red one. She had then forgotten to clean the original white blouse and put it back on when these later angles were shot. For whatever reason, no one on the crew had realized it either.

When the coverage of this scene is edited together later and the editor cuts from the wide shot to, say, the medium close-up, the supervisor's blouse will appear to change colors. This visual break in physical continuity will be so distracting that either the master scene or the subsequent coverage will be unusable.

Instead of a piece of wardrobe, it might have been a simple hand-position problem. Suppose the representative had her right hand curled under her chin in the medium close-up, but it was on her lap in the other angles. Later, in editing, every time a cut is made to this shot, her hand will appear to instantly jump to her chin.

These examples are only two of countless physical continuity problems that can be sources of distraction in and even ridicule of a director's work. When audience members become aware of physical continuity problems, the director's illusion is broken and the distraction may become the butt of audience jokes.

During production, the director should focus primarily on the performer—*whether it's a professional actor, an employee, or a company executive.*

—Patti Ryan
Producer/Director

The corporate director can avoid physical continuity problems in three ways. One is by keeping a close personal eye on all sources of physical continuity during shooting. This includes how the actors are positioned, what they are wearing, their physical actions in relation to dialogue in each scene, and basically all other elements within the camera's frame. The problem with this solution is that it can keep the director so busy that her attention may be distracted from other aesthetic issues, such as performance and composition.

The second solution is to assign the responsibility for physical continuity to someone dependable, such as the AD or a person taking script notes. In this way, the director can focus her attention on the other elements just mentioned and feel secure that someone with a keen eye for detail is watching physical continuity. "Keen eye," incidentally, is the critical phrase here; the person who watches continuity must realize its importance and maintain a strict vigilance throughout the shoot.

A third possibility is to review some visual reference of the scene. This might be video playback or Polaroid pictures of actors and sets taken by the AD at key points. For instance, in the "mismatching blouse" situation just mentioned, if both actors had had Polaroids taken of them just before lunch, or if the director had watched playback of the master scene immediately following lunch, the mismatch would probably have been discovered before shooting continued.

In feature film production, "continuity," as it's called, is usually a job in itself or a part of the script supervisor's responsibilities. In corporate television, when both the director and someone like the AD or a trusted PA work together to watch continuity, most times problems can be avoided.

Performance Continuity

Maintaining performance continuity involves focusing the same attention on similarity, but in a *non*physical sense. Consider these two angles in the same "host" scene:

FADE-IN:
EXT PARK—DAY
A company picnic is in full swing. There are tables and umbrellas; badminton games, as well as horseshoes and croquet, are in progress. The HOST steps into the frame and looks first over his shoulder and then at the CAMERA.

HOST
Well, here it is, at last—picnic day! The games are under way, the hamburgers are on the grill, and everyone's having a great time. C'mon, let's have a look around.

Host steps into ANOTHER ANGLE. We see that he is standing beside a table. Members of the Johnson family are laying out paper plates, plastic forks, cups, and napkins. Mrs. Johnson pours soft drinks.

HOST
Here at the Johnsons' table, it's strictly soft drinks. No alcohol—even for the adults.

Host exits.

Imagine the host entering the first angle in this script with a smile on his face and an energized sense of excitement in his delivery. This would certainly seem appropriate for the picnic setting.

But suppose when he stepped into the second angle, the smile was gone. Assume too that his tone had suddenly become serious or subdued. The result would be a break in performance continuity that would be extremely jarring to the audience.

This might seem like something the director would simply not let happen, but remember, just as with our previous scene, the second angle of this shot might not be recorded for several hours after the first one—

maybe not until after a large, heavy lunch and 30 minutes of napping in the shade. If this were the case, unless someone had remained keenly aware of the tone and energy level the host had used in the first angle, the second angle could easily be recorded differently.

The responsibility for performance continuity is the director's. To maintain it, she watches and judges each scene carefully, committing what she has seen to memory. Then later, when a second angle is shot, she must recall that first performance and compare the two in her mind, making sure the proper continuity is present. In the case of video, she may also watch playback of the previous scene to refresh her memory. As an additional bit of "insurance," she might also show playback to the actor as they discuss the proper energy level of the upcoming shot.

These same precautionary steps, incidentally, apply to audio recording. Many times, performance continuity is broken between on-camera and voice-over segments recorded at different times in different places. Working from memory and playback and enlisting the help of her audio recordist to match microphone and location qualities, the director must take careful precautions to ensure matching tone and energy levels between all recorded segments.

As you might guess, physical and performance continuity problems become increasingly complex and difficult to manage with longer schedules and out-of-sequence shooting. The experienced director is aware of this and takes the necessary steps in preproduction to ensure that her continuity remains intact throughout the shoot.

Screen Direction

Though screen direction problems are not often thought of as continuity faults, this is exactly what they are. In the case of screen direction, however, the lack of continuity relates to the *logical sense of motion* on the screen instead of performance or physical mismatches.

To understand the importance of this logical sense of motion, keep in mind that the viewer of a film or videotape develops a subconscious sense of relationship between people and objects moving on the screen. This relationship is often described in terms of something called screen direction.

For instance, if a scene establishes a train going from one city to another and the initial direction of travel is screen left to right, the audience will subconsciously expect to see the train traveling in that same left-to-right direction anytime it is shown en route. If we suddenly cut to a shot of it going in the opposite direction—screen right to left—a momentary disorientation occurs for the viewer as he reestablishes in his mind that the train is still indeed headed for where it had originally set out.

The same visual relationship is developed between people in terms of their movements, gestures, focus of attention, and dialogue. For instance, if two people are having a conversation and the director chooses to intercut back and forth between medium close-ups of each person, her audience will expect them to be facing toward each other. If one person is facing screen left to right, the other must be facing right to left, as

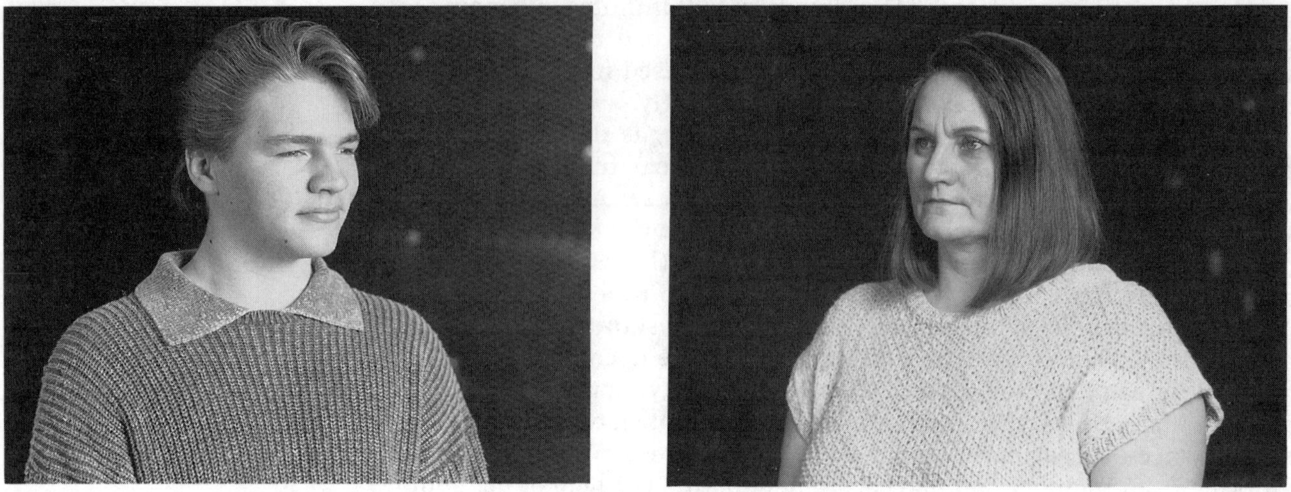

A B

Figure 8.1 Two medium close-ups facing properly.

shown in Figure 8.1. This will give the appearance that both people are facing properly. If one actor is shot in such a way that he faces the same screen direction as the person opposite him, he will appear to be looking away from the other person, as shown in Figure 8.2.

Again, this creates an immediate sense of disorientation in the mind of the viewer. He will subconsciously wonder something like, "Wait a minute. Is he still talking to the same person? Why did he turn away? Or did someone move?"

These types of screen direction discontinuities occur due to camera placement on the wrong side of what is called "the line," "the stage line," or "the screen direction line." "The line," as it's most often called, creates a fluid, imaginary, vertical plane through objects moving in any shot. In order to maintain consistent screen direction in her scene, a director cannot cross this line with the camera except under certain circumstances and only with great care given to how the resulting footage will cut together.

To illustrate, let's imagine shooting a very simple sequence involving several elements.

EXT HOUSE, IN THE DRIVEWAY—DAY

An employee is seated in his truck preparing to leave. A customer is leaning in at the driver's side window.

CUSTOMER

Thanks a lot. Nice work.

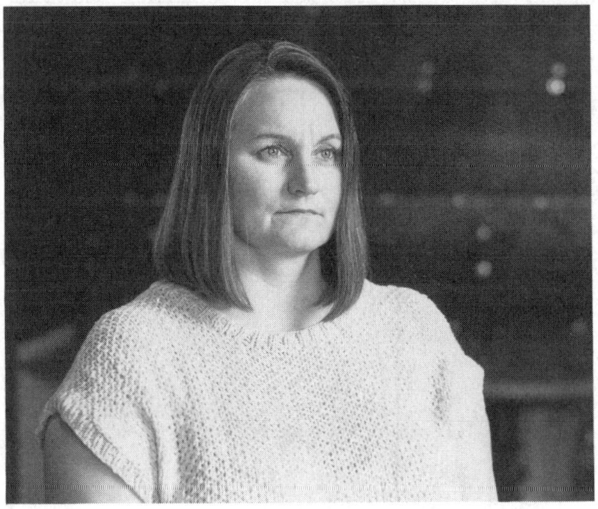

A

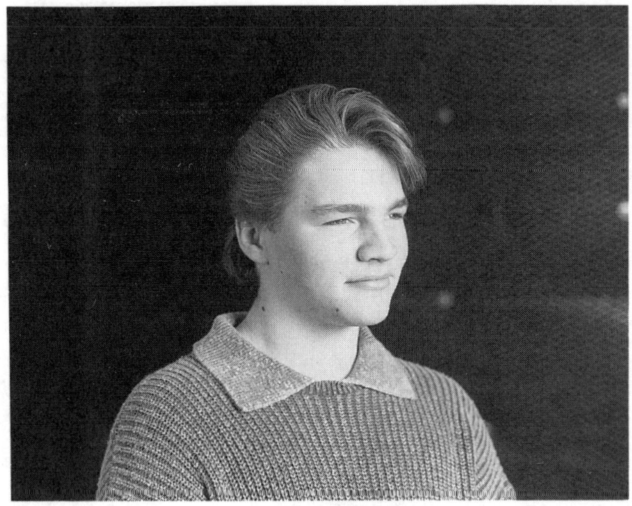

B

Figure 8.2 Crossing "the line" creates the illusion that one person (in this case, the young man) is facing away.

EMPLOYEE
Don't mention it. Call again if you run into any more problems.

CUSTOMER
You bet. Drive carefully, now.

The employee and the customer shake hands. As the customer walks away, the employee starts the truck, backs out into the street, and pulls away.

As we first begin to view this scene, "the line" is established, as shown in Figure 8.3. As long as the director shoots on the "A" side of this line, the eye lines of the customer and the employee will keep a consistent and properly established screen direction. The customer will always be facing screen right to left, and the employee will always be facing screen left to right.

But suppose the director has the DP move the camera to the "B" side of the line and shoot a close-up of the customer. If this shot is now cut into the conversation sequence, a reversal of screen direction will occur creating an immediate disorientation for the viewer. The customer will suddenly appear to be facing away from the employee.

As we have said, however, "the line" is also fluid. This means that it changes depending on the center of focus of the shot, as well as on the objects and motion involved. For example, in the second part of the scene, the employee backs his truck out onto the street. As soon as the customer

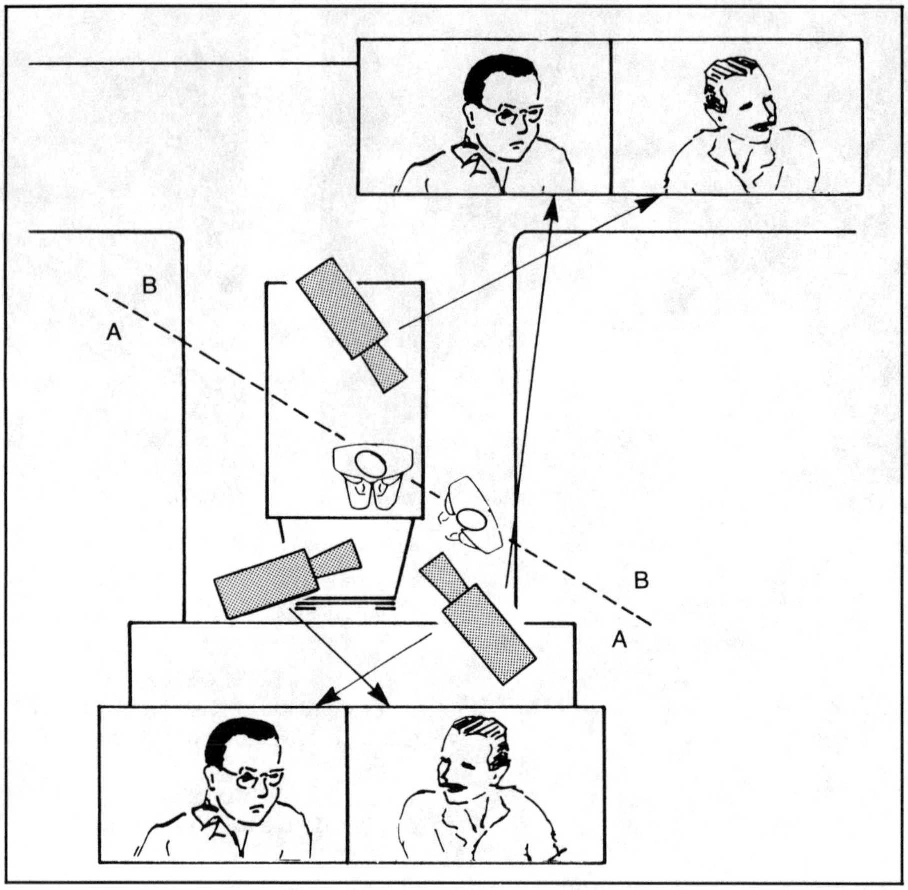

Figure 8.3 "The line" passes vertically through an overhead perspective of customer (right) and employee (left) in truck. Addition of camera position on B side crosses "the line," creating a close-up that faces the wrong way.

leaves and this backing action becomes the focus of attention, "the line" changes to a plane running down the center of the truck, as shown in Figure 8.4. This plane moves with the truck as its direction of travel changes.

To illustrate, we could set up a series of new camera placements and consider how the various shots would cut together.

As shown in Figure 8.5, the truck backs out of the driveway, moving away from the camera at position 1. It then starts down the street. At this point, it is moving screen *right to left.*

If the second angle of this coverage is shot from position 2, which way will the truck then be moving in the frame? If you said screen *left to right,* you're correct. You can see, then, that when these two angles are cut together, it would appear that the truck had started off one way and then suddenly reversed its direction.

Let's now consider the same action and set up two additional camera positions. In Figure 8.6, camera position 1 remains on the driver's side. Position 3 is along the same direction of travel, but on the passenger side. Position 2 is across the street in the same location as in the first setup, and position 4 is directly across the street from position 2.

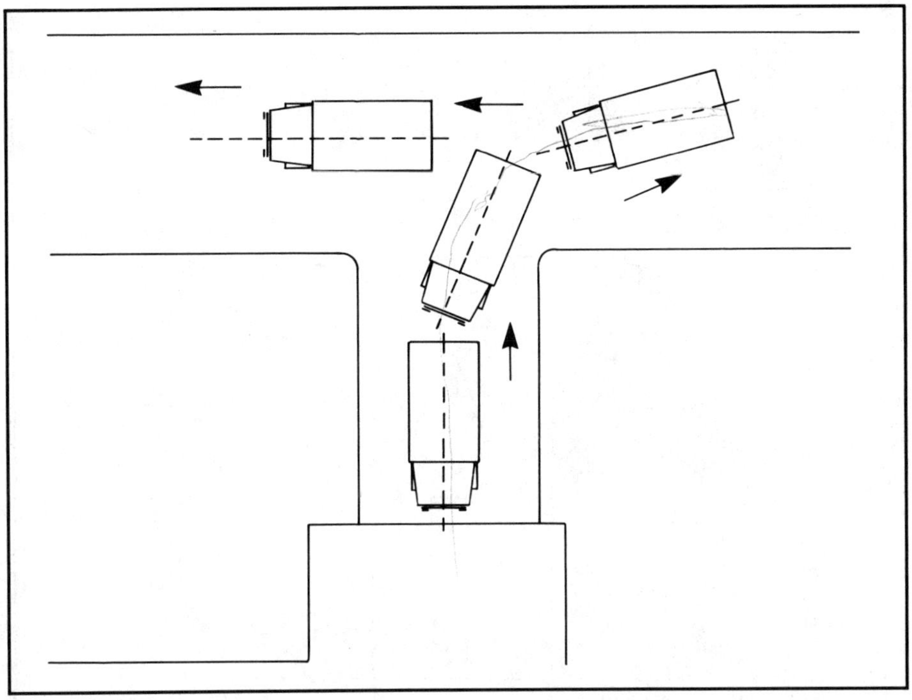

Figure 8.4 When focus of attention changes after customer leaves and truck begins to back out, the line changes also. Direction of movement is now focused on truck.

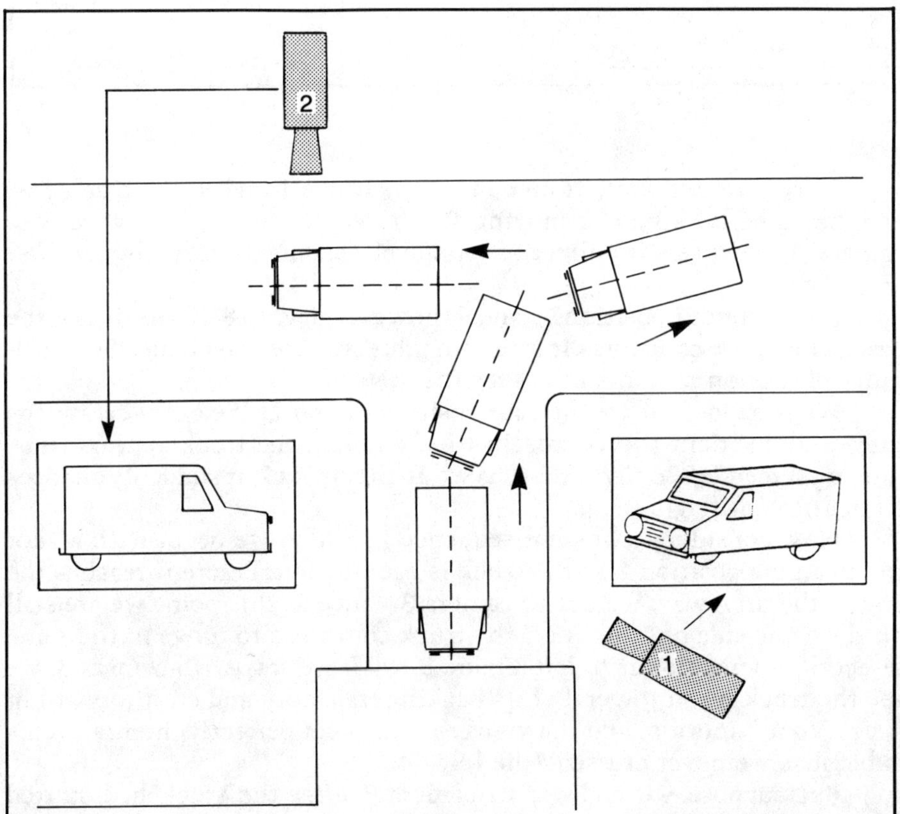

Figure 8.5 Backing truck shot from opposite sides of "the line" (positions 1 and 2) will reverse the screen direction, disorienting the viewer.

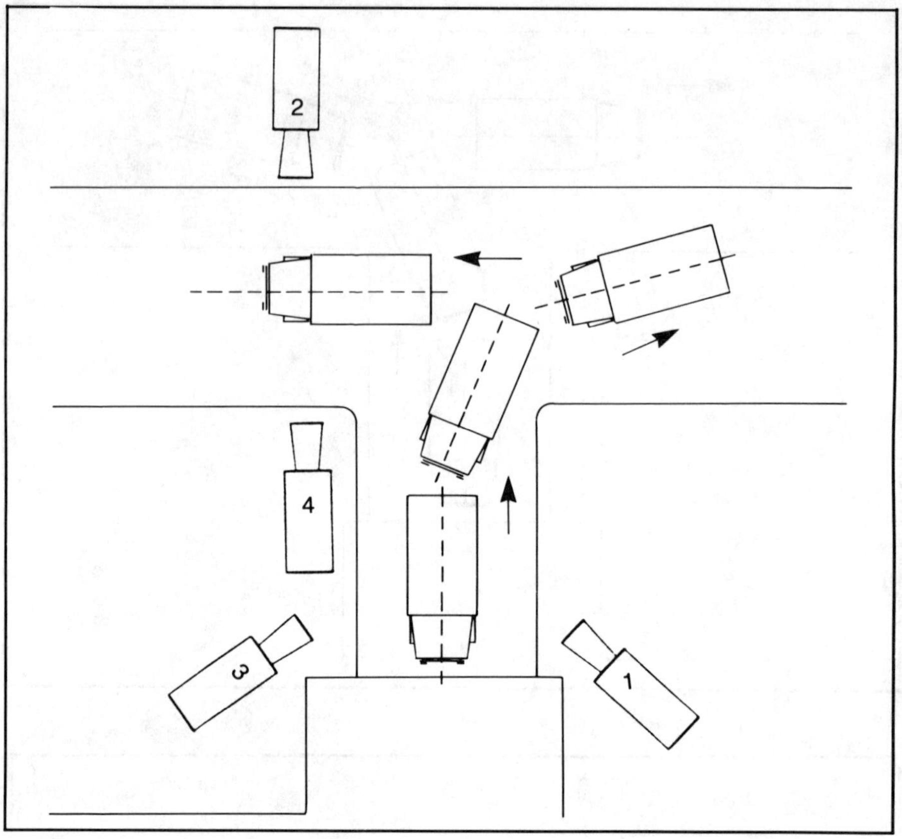

Figure 8.6 Several new camera positions. If shots from each are cut together at different times during the backing action, they will have various effects on screen direction continuity.

If we intercut shots from camera positions 1 and 3, the truck first appears to be backing out moving screen *left to right.* When we cut to camera 3, however, its direction of travel becomes screen *right to left* (Figure 8.7).

If we intercut positions 1 and 4, however (Figure 8.8), this is not the case. The truck continues in the same direction of travel and the continuity of motion remains intact for the viewer.

What happens if we intercut positions 1 and 2? We again cross "the line" and disorient the viewer. In the first shot, the truck appears to be moving screen *left to right;* then, as we go to camera 2, it suddenly changes to *right to left* (Figure 8.9).

Now consider a two-shot sequence in which we begin with a shot from camera position 2 as the truck is backing out. Before it reaches the end of the driveway, we cut to camera 3. Since at this point we are still on the same side of "the line," the truck continues to move in the same direction—screen *right to left* (Figure 8.10). If we stay with camera 3, we see the truck reach the end of its backing trajectory and start forward in a *right to left* motion, and the viewer is still kept perfectly oriented. This is because we never crossed "the line."

But suppose we had cut to camera 3 *after* the truck had started forward. At this point in the motion, the camera positions are on opposite

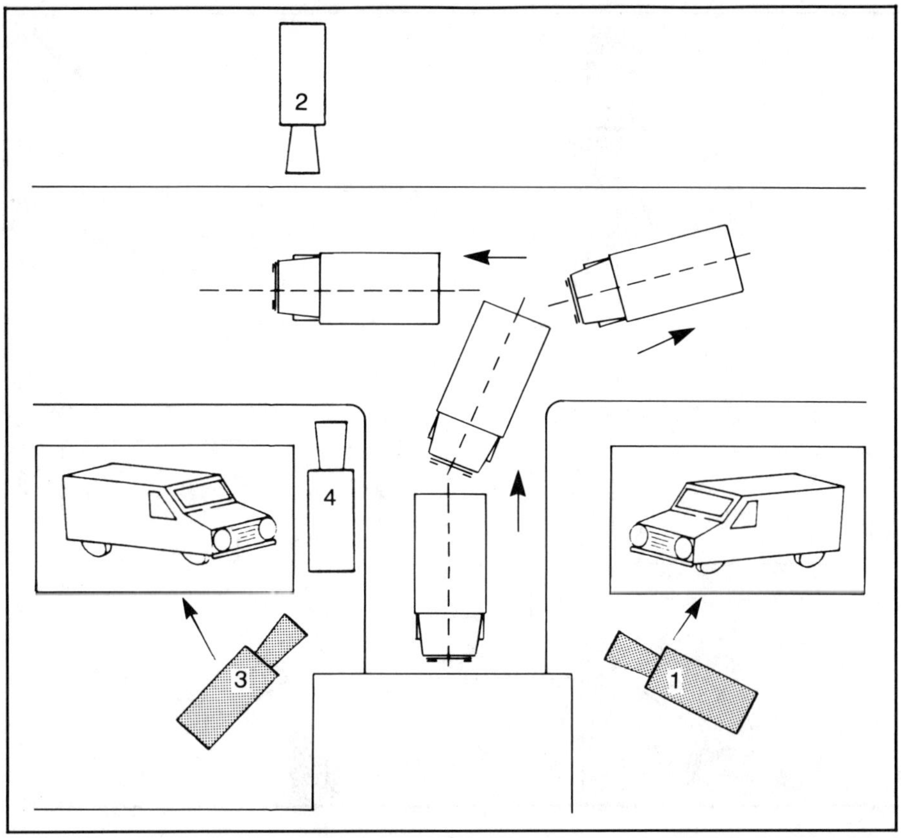

Figure 8.7 These angles (positions 1 and 3) will not work visually, since "the line" is crossed.

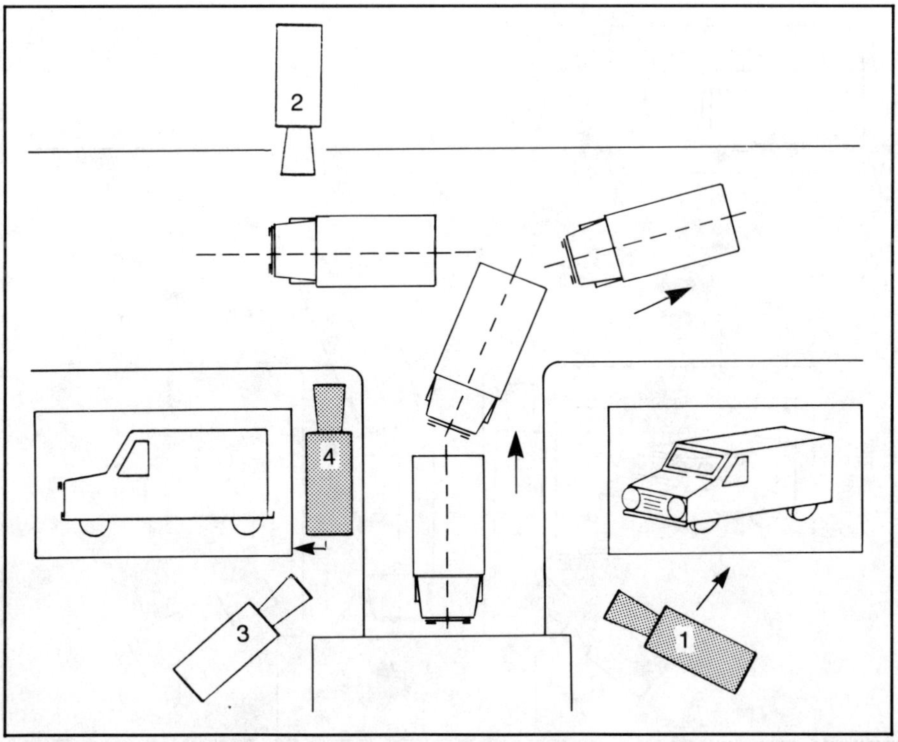

Figure 8.8 These angles (positions 1 and 4) will keep the continuity of motion intact.

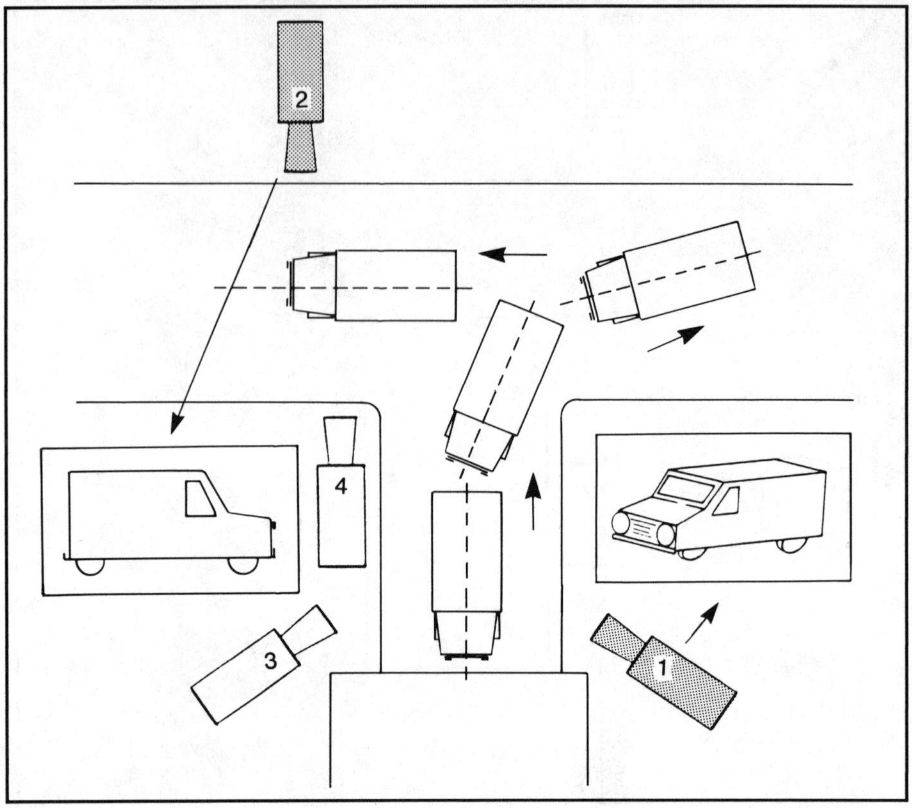

Figure 8.9 Positions 1 and 2. Again, a screen direction reversal.

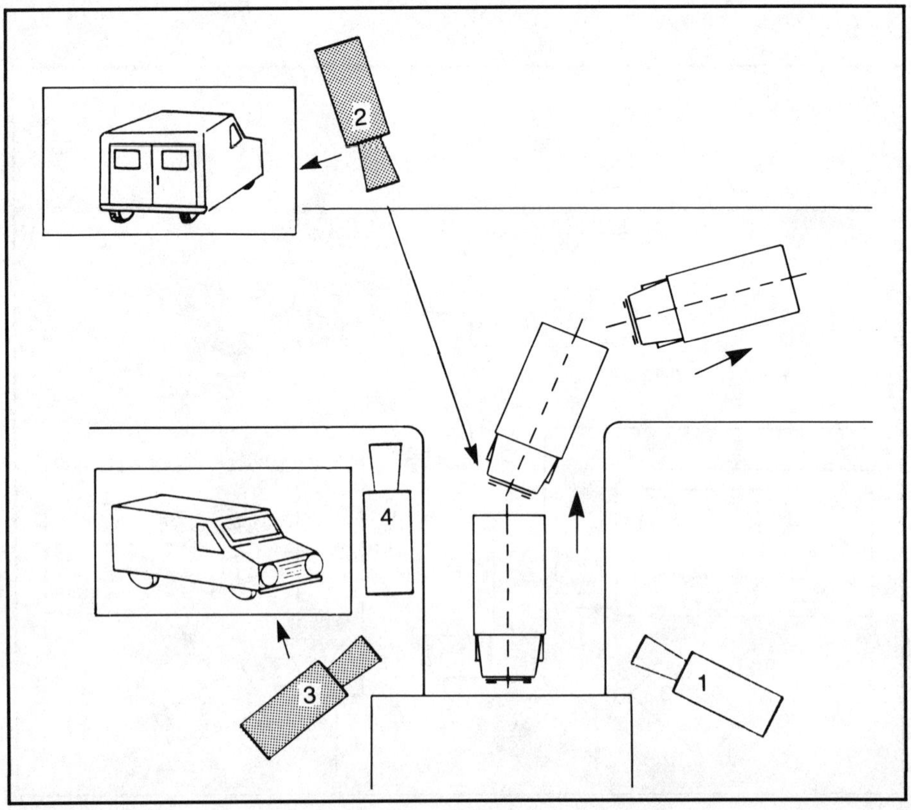

Figure 8.10 Screen direction will remain proper as long as the cut to camera 3 is made *before* the truck reaches the end of the driveway. After this point positions 2 and 3 are on opposite sides of "the line."

sides of "the line" and so a reversal of screen direction would have occurred (Figure 8.11).

Thus, you can see that the stage line creates an imaginary plane running down the *center of motion* of any object, and as that center of motion changes direction, so too does the position "of line."

Whether it's objects or people, the director must remain aware that "the line" exists in every shot. Constant attention must be paid to ensure that the people and objects in those shots are moving logically and consistently.

"Sneaking" Across the Line

With so much talk of staying on one side of the line, you might wonder if it is acceptable to ever cross the line. The answer is "yes," and the rule of thumb is simply to be sure the change is not disorienting to the viewer.

To illustrate, let's return to our truck backing out of the driveway. In Figure 8.6, we established that camera positions 1 and 3 were on opposite sides of "the line" and that intercutting between them would cause a discontinuity of motion for the viewer. This assumes that both camera positions are nearly perpendicular to the truck's backing motion and that the focal lengths and levels are roughly the same.

If we move the cameras to the positions show in Figure 8.12, how-

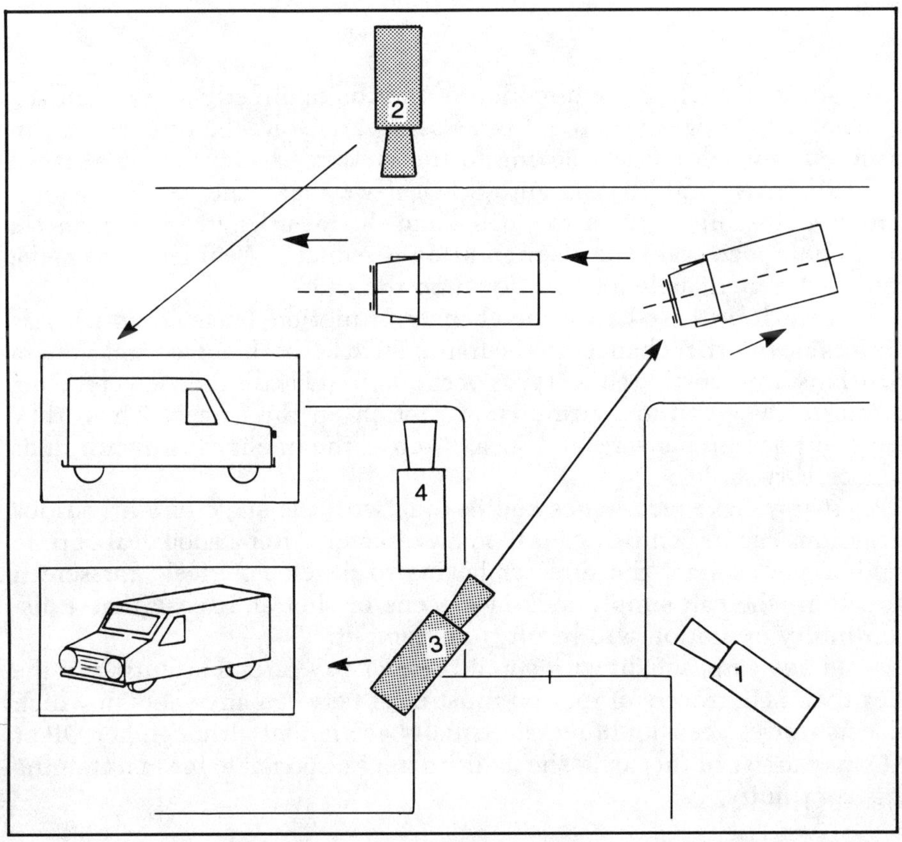

Figure 8.11 Reversal.

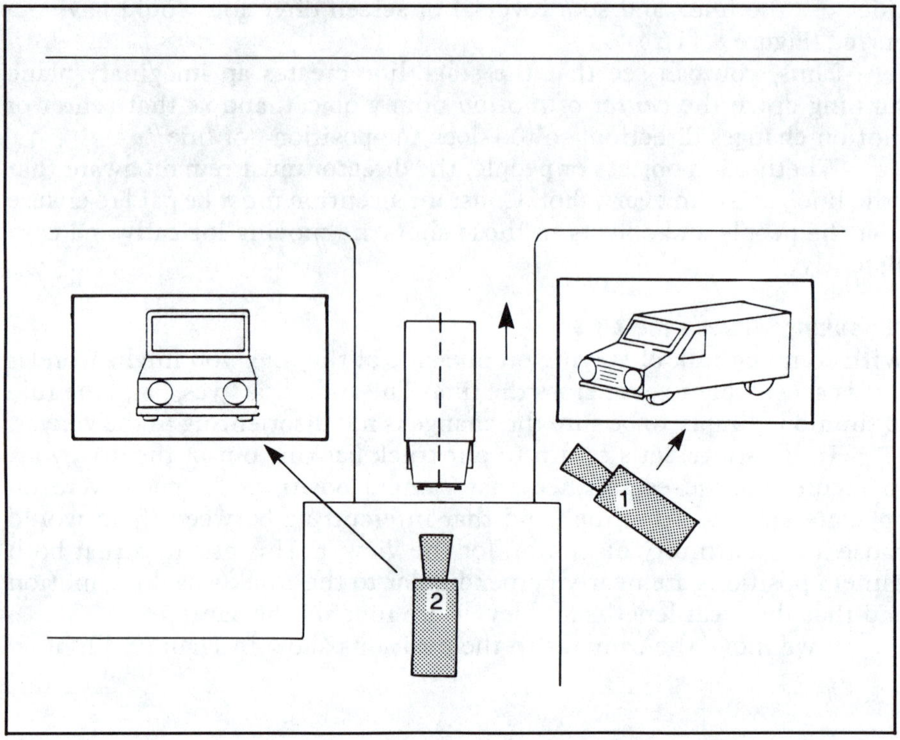

Figure 8.12 Cutting from an angle *on* "the line" (camera position 2) to one side or the other will not be visually disorienting.

ever, since in angle 2 the direction of motion is directly away from the camera, the change of motion from a perspective *on* "the line" to one on either *side* would be less jarring to the viewer. In addition, even if we actually *cross* "the line" (Figure 8.13), if we make the camera 2 focal length a close-up—say, of the grill—and the focal length of camera 1 a wide shot, once again the disorientation is reduced. We might also make camera 1 a high angle looking down on the truck.

In each of these cases, the changes of motion are made much less noticeable by other changes in the frame, such as focal lengths and camera positions. Intercutting these types of shots would leave the viewer feeling comfortable with the continuity of motion on the screen. Thus, they represent acceptable ways to "sneak" across the line if the director finds it necessary to do so.

Using these techniques and dealing with the stage line in various situations can be confusing for the new director. After a good deal of time behind the camera, the director begins to develop a "feel" for screen direction; she can simply look at a scene or shot and sense that a discontinuity of motion will result from using it.

In any case, the director should always pay careful attention to the idea that a continuity of motion must exist between any shots in which objects move. She should remain equally aware that although her DP or AD may assist in this area, she is ultimately responsible for maintaining this continuity.

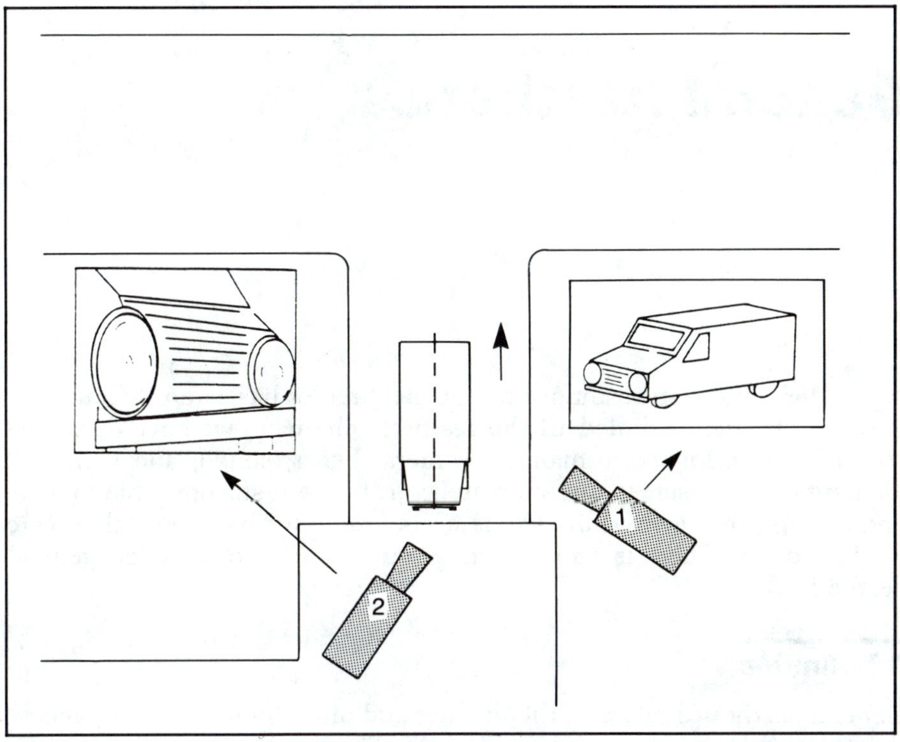

Figure 8.13 "Sneaking" across "the line" is done often. If shot properly, it works fine.

Summary

As you can see, avoiding problems that negatively affect the technical aesthetics of a director's work is also extremely important to a credible final illusion on the screen. As a new director, you should become very familiar with the ramifications of technical and continuity elements of your program and continually strive to achieve a sense of technical aesthetic perfection.

9 General Aesthetics

Again, let's make an assumption that the director has carefully analyzed and appropriately handled all the aesthetic elements we have discussed so far—the script, performances, camera, lights, sound, and technical elements such as screen direction and continuity. Is she now able to relax and assume that everything she is accountable for has been taken care of? Not quite. There is still one more critical area to consider: general aesthetics.

A Definition

General aesthetics refers to the broader and often more subtle aspects of the program: *tone, pace,* and *transitions.* The subtle nature of these elements does not reduce their importance. Like so many other subtleties under a director's control, these can have a profound impact on the aesthetic quality of any show.

Although we have previously discussed tone and pace in terms of script aesthetics, it's worth our time to re-examine their importance as they relate to production.

Tone

You may recall that words that describe program tone are *positive, negative, inspirational, informational, sad, happy, uplifting, depressing,* and so on. Although a program may have a single *overall* tone, it may have multiple tones throughout.

For instance, a program that explores the process of striving to achieve one's aspirations might begin with an *informational* tone as a host sets the stage by explaining the subject. The same program might then follow with a series of *sad* or *negative* scenes as a main character is seen trying to excel but continually failing at what he attempts. These might lead to a series of additional *informational* scenes as the host explains what the character must do to succeed and eventually follow with several *positive* scenes in which the character takes the proper steps and finally makes good. The end result of all these might be an *overall* program tone viewers would call *uplifting.*

As mentioned previously, during her initial script analysis, the director carefully considers these types of tonal elements. She then strives to create the tone the writer has intended and the producer and client have approved. In production, this translates into decisions about such things as lighting, wardrobe, camera positions and so on.

For instance, the first host scenes might take place on a dark limbo set with random light patterns thrown on a lattice or cyclorama in the background. While this would maintain the informational tone needed to start the program, it would also hint that the subject matter was serious, thus setting the stage for the sad or negative scenes. These scenes might also be lit darkly with an accent on shadows. The character's makeup and wardrobe might reflect frustration and sadness. His clothes could be rumpled and askew, and perhaps he could be made up to appear tired. Camera positions could also be adjusted to help achieve the tone of sadness or frustration by being slightly tilted or oblique in perspective.

When the host is seen again, if what he says begins the sequence that changes the tone, he might appear in a new position on the set or under lighting that is brighter and more upbeat. Perhaps the initial dark shadows could be brightened and softened. These changes could also be accented by the host's performance, which might be more positive and upbeat.

Later, when the character is seen achieving his goals, he could appear fresher and more positive in his dress and makeup. The lighting, camera positions, and, of course, the actor's performance could also reflect this change.

The sum total of these many individual tonal elements would help create the final, overall tone mentioned earlier—a sense that the story was uplifting.

Corporate Tone Considerations

In many cases, the general tone a corporate program is expected to achieve is *positive.* This is because executives and corporate clients often feel it is critical to stress the positive aspects of any subject they wish to communicate.

This can sometimes create a dilemma for the corporate director, if the subject matter does not appear to warrant a positive treatment. Most times, however, it is the writer who must fight these battles as the script is being developed. By the time the director is brought in on the project, the written scenes reflect what the client and producer have agreed will work. It then becomes the director's job to deliver the scenes as scripted. While this may seem simple enough, it can sometimes become difficult due to the subtle nature of tonal elements.

I know of one director who was given a script with what could have been called a "serious informational" tone. It consisted of a tour by an on-camera host through a typical residential home. While on this tour, the host discussed the dangerous chemicals that many of us unknowingly keep in our homes. He also talked about ways to handle and dispose of them.

To give the program more interest and add what he considered a "light touch," the director decided at the last minute to include a youngster in many of these scenes. As the scenes were shot, the host would occasionally pass the boy during the tour and find him in humorous situations—eating cake in the kitchen, wearing headphones, playing loud rock music in the den, leaving a mess in the garage, and so on.

The addition of the youngster succeeded in doing exactly what the director had wanted: It added a light touch. But it also changed the tone of the program. Instead of informational *serious*, the program became informational *light*. This was something the producer had not wanted, considering the seriousness of the subject matter. As a result, the youngster's scenes were cut before the program was completed.

This is just one of many possibilities, but the caution is always the same. Be conscious of the tonal elements of any program you direct, and consider how you can *support* or *enhance* what the script strives to achieve—but do not *change* it without the full prior knowledge and approval of both the producer and the client.

Pace

Pace, as we have said, can be thought of in terms of "life speeds." These speeds are manifested on the screen through three elements: *shot content*, *shot length*, and *editing*.

Shot Content

Whatever is contained in a shot has pace. If actors are seen moving carefully and methodically, the pace will be slow. If actors are seen rushing about, the pace will be fast. These same shot content rules hold true for dialogue and narration.

Consider the following scene:

EXT CREEK—DAY

After a long day and a huge barbecued steak dinner, Anne and Jean are lying in their sleeping bags near the camp fire. Both are staring up at the stars.

ANNE

Have you ever had a more fun day?

JEAN

(a whisper)

If so, I sure can't remember it.

ANNE

Me either. But I'm so tired I don't think
I'll be able to keep my eyes open long
enough to enjoy all these stars. . . .
Look, there's the Big Dipper . . . see it?
Up there through the branches? Jean? . . .

Anne looks over and finds that Jean is asleep.

DISSOLVE

It's obvious that a slow, relaxed pace is called for here, in terms of both movement and dialogue delivery. If these actors were fidgeting about nervously or the lines in this scene were delivered with a rapid pace, it would ruin the tone and thus the credibility of the scene.

Shot Length

By necessity, shots that are short in length dictate rapid editing. This rapid editing will lead to a more energetic pace. To illustrate, imagine that the camp-fire scene we just described is being shot by two directors.

The first director starts by shooting a wide establishing shot in which we see the two sleeping bags lying on a forest floor with the camp fire nearby. Next, he moves the camera in for a two shot in which the perspective is looking down on both Anne and Jean. From this position, he records all dialogue. Next, he shoots a close-up of Anne only, performing all dialogue and following this, a close-up of Jean. Thus, he has covered the scene with four shots.

The second director shoots only one shot. He mounts the camera above the two women and begins with a shot of the camp fire alone. After several seconds, he begins a slow pan revealing Anne and Jean in their sleeping bags. As Jean begins to talk, the camera zooms in or lowers slowly, moving first into a two shot as the women exchange their first lines and finally to a close-up of Anne's face alone. This camera motion ends just as Anne is about to look over at Jean. As she does so, the camera pans a final time to reveal that Jean is asleep.

To see what effect these two shooting styles would have on the finished piece, we must consider how each will be handled in editing.

Editing

Editing affects the pace of a program primarily in terms of the *frequency* of edits. Short, rapid cuts typically create a jumpy sense of speed and excitement or tension. Long takes with infrequent cuts, however, often create a slower, smoother pace.

In the coverage shot by our two directors, you will recall that the first director chose to cover the scene with a series of short static shots. His version of the scene might then be edited as follows: The establishing shot of the fire is used for a few seconds of silence, until Anne begins her first line. As this line begins, cut to the two shot to establish both women. As Jean says, "If so, I sure can't remember it," cut to her close-up. As Anne speaks again, cut to her close-up. As Anne looks over and sees that Jean has fallen asleep, cut to Jean's close-up showing her eyes closed. Finally, cut back to the establishing shot before dissolving to the next scene.

The first version of this scene thus cuts together five short segments. Although the editing pace this creates could certainly work, it is somewhat inconsistent with the smooth, relaxed tone the scene should be communicating.

The second director's coverage of the scene consisted of one long, slow moving shot that would require only a single edit. This version accents the calm, relaxed tone of the scene and thus enhances the overall illusion.

As mentioned earlier, tone is a subtle aesthetic element. This means that it is often affected by other elements, such as actors' performances, lighting, sets, and props. Additional considerations are the production schedule and budget. It can sometimes take more time and equipment to get a long single shot than it takes to cover a scene with several standard shots. In short, hard-and-fast rules about tone are difficult to establish. Every scene is different, and it is up to the director to create the proper tone by shooting her scenes effectively. The experienced director realizes this and analyzes all production elements when considering the type of coverage that will be required.

Transitions

With the tone and pace of a program established in the director's mind, she must also consider the transitions *between* her shots and scenes. This too involves editing.

There are four primary edits used to provide transitions from one shot or scene to another: *cuts, dissolves, wipes,* and *digital video effects (DVE).* Since each of these can also have important effects on the general aesthetic quality of the program, it's worth taking the time to briefly discuss each.

Use of the Cut

A cut is the simplest way to get from one picture to another. A cut simply means that one picture ends and another begins immediately. In most cases, a cut is the preferred transition, since it makes no "statement" about the content of the program. A cut is usually meant to happen without audience recognition; thus, it lets the pictures and sound carry the complete message.

This is the case, for instance, when a conversation in a sit-com cuts from one person to another. As viewers, we are unaware of these cuts. They happen subliminally and seem to coincidentally transport us to the most logical viewing perspectives as they become appropriate. This same subliminal quality is in effect when cuts are used as transitions between *time frames, locations,* and *general subjects.*

In some cases, however, a director can use a cut to make a statement. Instead of going unnoticed, it can punctuate a transition. Consider this scene:

INT OFFICE—DAY
Lots of people, business, and office SOUNDS. Janice is rounding the corner by her desk. Her foot catches on an electrical cord. She trips and falls toward another, close-by desk. On the desk is a large, acrylic square decorative piece. Janice's face is headed straight toward a pointed corner of it.

JANICE's POV—The CAMERA ACCELERATES straight toward the corner of the piece, and suddenly—

CUT TO BLACK/SILENCE

In this case, the cut to black and silence at the moment of Janice's impact with the sharp corner heightens our awareness of the transition. In doing so, it drives home the point—both visually, by going to black, and through the use of sound (in this case, a lack of it)—that Janice is, at best, left unconscious. The previous cut, however, from our view of Janice to her point of view (POV) would happen with no awareness on our part.

Use of Dissolves

A dissolve overlaps one picture onto the next during the transition. For this reason, it is a slower and usually softer way to transition from one picture to the next.

Dissolves often do make statements. They tell the viewer that time has passed or some other major change has taken place. Sometimes they say both. For instance:

EXT HIGHWAY—DAY
The company truck moves off on a long, flat, empty, country highway, toward a setting sun. MUSIC UP—A RAMBLING PIECE.

DISSOLVE

EXT MOUNTAIN ROAD—DAY
The truck rounds a corner on a steep, snowy mountain road.

DISSOLVE

EXT OCEAN HIGHWAY—DAY
The truck winds along the sunny Pacific Coast Highway.

DISSOLVE

EXT FREEWAY—DAY
CAMERA starts on a shot of a freeway sign for the exit to Hollywood Freeway. As the shot widens, the company truck moves into the frame. With its turn signal on, it heads toward Hollywood. MUSIC CONTINUES.

DISSOLVE

EXT WAREHOUSE—NIGHT
Starting on a shot of the "Benson Furniture Warehouse" sign, CAMERA WIDENS to reveal the company truck pulling up in front. As it comes to a stop, MUSIC STING.

In this case, the dissolves help tell us that time is passing and so are major locations. Supported by the rambling music piece, they help carry us across country—again, totally unaware of the camera or the transitions from one scene to another.

The length of a dissolve can also have varying effects on the segments it is used to bridge. A very slow dissolve, for instance, can make one picture appear to be superimposed over the next. This can create an eerie, or dreamlike effect. Faster dissolves, sometime referred to as "soft cuts," can be used to take the hard edge off an otherwise rough or jarring cut.

Use of Wipes

Wipes cause one picture to "wipe" off the screen as another appears. This is usually done with the vertical edge of one picture traveling from right to left or vice versa across the screen. Wipes can also be made in circles, squares, or various other shapes and patterns. Like dissolves, wipes usually signify a passage of time or a major change of some sort. They also call immediate attention to themselves, however, and thus should be used sparingly.

The following scene is one example in which a wipe might be used.

INT WIZARD'S CASTLE—NIGHT
Merlin looks into the crystal ball and smiles.

MERLIN'S POV—The ball. Deep in its glass, something is moving.
From the ball . . .

CIRCLE WIPE OUT TO

INT WITCH'S CAVE—NIGHT
Ingrid stands over the boiling caldron. Surrounding her, animal bones and skins hang on the dark dirt walls. Suddenly, John leaps out out of the caldron, and says . . .

JOHN
Nice place you got here! Kind of, ah, earthy and . . . Say! I like your taste in wall treatments!

A circle wipe would probably work in this situation for three reasons. First, it is used to suggest a major change of location. Second, it would punctuate the transition from the POV shot of the circular-shaped crystal ball. Third, and perhaps most important, it would enhance the humorous tone the writer strove for. In this type of tongue-in-cheek application, if an effect such as a wipe calls attention to itself, so much the better. In a similar scene with a serious tone, a cut or dissolve would probably be more appropriate.

Use of Digital Video Effects (DVE)

DVE, as it's often called, can do virtually anything to a video picture. It is DVE equipment that makes a picture curl up like a piece of paper, turn into a ball and bounce off the screen, become one side of a spinning box, or do any number of other, often very flashy visual effects.

Digital effects call attention to themselves on purpose. They are often used in corporate magazine-type shows and marketing programs that require a high-tech production look. They are also used many times simply to give corporate programs an expensive, broadcast appearance. Producers reason that today's typical corporate viewers are very visually sophisticated because of what they see continually on their TVs at home. If corporate programs don't match up to this level, producers reason, they will be considered inferior. While there's undoubtedly some truth to this argument, it should not be considered a license for unrestricted DVE use. There are also many other ingredients that affect thc quality of a corporate show—primarily good scriptwriting, credible performances, and audio-visual aesthetics.

Summary

We've found that the tone, pace, and transitions in a corporate program, like so many other subtleties a director must deal with, are critical to the show's illusory quality. The wise director remains aware of this fact throughout the entire production process and strives, as with all other aesthetic elements, to keep them totally credible to the viewing audience.

10 People Skills

❑ ❑ ❑ ❑ ❑

The director's job goes beyond being creative. You are the team leader *of the production crew.*

—Patti Ryan
Producer/Director

One of a director's attributes might be called aesthetic sensitivity. This sensitivity enables a director to consider all of the aesthetic elements we have been discussing and make basic judgments about their credibility or aesthetic quality. When all is said and done, however, no matter how much aesthetic sensitivity a director may have, he cannot achieve his goals of creating and maintaining his media illusion without a strong ability to work with *people.*

His actors must deliver sincere performances. His DP must light with speed, skill, and creativity. His AD must take accurate script notes and remain alert to possible continuity problems. His sound recordist must constantly monitor audio quality and levels. His grips and gaffers must hustle to keep the shoot on schedule. His client must support and approve his work, as must his producer.

For these reasons, good people skills are not just important but an inherent part of any director's role. They allow the director to act effectively in four basic capacities: as *consultant, manager, supervisor,* and *coach.*

The Director as Consultant

When first brought in on a project, the director is a consultant to the producer and the client. He advises both on how the script will manifest itself in visual form, and he suggests possible changes if he feels they will improve the end product. Two key words in this statement are *advises* and *suggests.* They are important because they leave the final decisions on script changes up to the producer and client.

Like that of any other consultant, the director's job in this early phase of preproduction is simply to make these two critical players aware of any aspects that could have a negative impact on the project. If, having heard the directors' opinions, the client or producer decides to go ahead with the script as is, the director should then set about doing exactly that to the best of his ability. The exception to this rule, of course, is a script that the director feels is simply too bad to attach to his name.

Some directors find this process of conceding to client and producer wishes difficult. If you feel you may fall into this category, the important thing to remember is that if you are hired to direct a corporate program, you are certainly involved in an important project, but you are not directing an epic. You are being paid to meet the *client's and producer's* requirements. Directors who become obstinate about making certain changes or overly possessive about their creative control usually end up becoming too much trouble for the corporate producer to deal with.

When a new project arises, rather than "fight" over changes she would like made, the producer will probably just hire another, more agreeable director. Or, if the friction has been between the client and director, there's a good chance the client will make it clear to the producer that she chooses not to work with that individual again. In most corporate environments, the producer will be obliged to go along with the client's wishes, whether she agrees or not.

To illustrate this point, consider the following scene:

EXT BACKYARD—DAY
The meter reader opens the gate and enters the yard. She starts toward the meter. After approximately five steps, she hears a noise and looks up. A huge German shepherd is standing in a far corner of the yard—teeth bared. It charges.

The meter reader dashes for the gate, "flies" through, and closes it just in time, as the shepherd slams into it.

In an effort to give real credibility to his work, a director may visualize a scene like this being shot with both the dog and the meter reader in the same shot. If his reasoning were that such a shot would heighten the scene's intensity and sense of reality, he would be correct.

The producer, however, may have a different perspective. She may feel that recording the shot as the director would like might either endanger the actor or, if a professional animal handler were used, be too expensive and time-consuming for the budget.

Her alternate suggestion might be to record the dog in one series of shots and the meter reader in a different series. The result could still be very credible if shot and edited properly, though perhaps not *quite* as intense as seeing the dog actually nipping at the meter reader's heels.

In this situation, having made his case for "reality," the wise director would concede to the producer's wishes. He would realize that the scene could play out effectively either way and that not getting exactly what he wanted would not be worth jeopardizing a professional relationship.

The *un*wise director might push the issue to the point of appearing to suggest that if the scene were not done his way, it simply would not be effective. Besides seeming obstinate, this suggestion carries with it an indirect statement that the producer is a poor judge of what should or shouldn't be recorded for her script. Making such a statement would make a poor example of the director's consultant skills—and one that would probably find him out looking for work.

The Director as Manager

A manager oversees a project, its people, or both. In the director's case, the word *both* is accurate. From the time he begins to prepare a project for production, he will be entrusted by the producer with managing *people* and *resources*.

The people include the entire production team, the cast, and perhaps the off-line editor. We'll discuss these relationships in more detail shortly. The resources include all other elements of the production, such as props, locations, permits, equipment, and schedules. Since each of these elements also involves money, the director is a manager of anywhere from a few thousand dollars on a very small shoot to $100,000 or more on a big-budget corporate shoot.

The director's managerial skills apply mostly during preproduction and deal with the general aspects of the production, such as *planning*, *organizing*, and *confirming* all aspects of a shoot.

Planning

The director plans by applying his specialized knowledge to a script and judging the time, people, and resources required to get the project "in the can." In order to accomplish this, he must first decide how he will shoot each scene. Will his needs require a dolly rental at $100 a day? Or can the camera record the action from a tripod? Will paid extras be required to move about in the foreground and background to create a sense of reality? Or will friends or employees perhaps be available? Will lights be required on location, making the rental of a generator necessary? What type of mics will be required, and do they involve the rental of special equipment or accessories?

These and many other similar questions must be answered by the director in preproduction for the production *plan* to be accurately formulated. That plan usually consists of the following documented items:

1. a script breakdown itemizing all elements, both human and otherwise, required to be recorded in all scenes of the shoot
2. a shooting schedule outlining the shoot scene by scene and hour by hour
3. a crew list showing all crew members and their roles, workdays, and "call times"
4. a shot list, which is a director's listing of the types of shots he plans to record to cover each scene
5. an equipment list showing all equipment needed for the shoot, the days it will be required, where it is to be obtained, and perhaps its total cost
6. a talent booking sheet listing all actors and actresses, either professionals or company employees, along with the dates and times they will be required
7. various other departmental documents required by the producer, client, or director

As we mentioned previously, though the AD or a PA may do much of the legwork on many of these items, the director is the guiding, approving, and thus *managing* force behind them.

Organizing

The director then *organizes* by bringing together all of the people and elements just mentioned and arranging them in a way that will allow everything and everyone to be utilized in the most cost-effective way. For instance, he will carefully choose locations considering costs, travel time, logistics, and other factors. He will then schedule the use of each location so as to get as much shooting accomplished as possible in the shortest time. He will consider crew and talent also, in terms of what job requirements are needed on what days. For instance, on the days with the dolly work, will an extra grip be needed to push the dolly, lay the track, or perhaps pull focus for the DP? Is makeup needed only on the days with principal talent? How many trucks must be rented or reserved on each day to accomplish this? And so on.

Confirming

Confirmation of all elements brought together in preproduction is extremely important, especially if the shoot is a large or complex one. It is one thing to plan and organize effectively, but it is another to be absolutely sure everything has been arranged and will arrive *when* it is supposed to, *where* it is supposed to, and *how* it is supposed to.

The feature film and prime-time television director has less to worry about in these areas, since she has staffs of people dedicated to nothing more than being certain that all loose ends are taken care of. In corporate television, however, such luxuries are rare. Usually these responsibilities fall squarely on the director's shoulders. And if they are not carefully attended to with great attention paid to even the smallest details, the shoot can be stopped short for hours.

For instance, a bolt that attaches a camera plate to a particular type of tripod head is essential to be able to shoot. Sometimes, however, such bolts may not be stored with the camera. The key to an office building may be the only possible way to get into a location on a weekend—*40* miles from the studio. If shooting will occur in a business lawn area, are there automatic sprinklers that may suddenly come on in the middle of a take? If so, does someone on the crew know how and where to shut them off?

Preproduction meetings It is easiest for the director to confirm the elements involved in his shoot by having one or more preproduction meetings with his AD and perhaps the entire crew. In such meetings, the director can simply go down his "To Do" list and get updates and confirmations on various aspects of the production as they develop.

One effective way to perform a final confirmation is to sit with the AD and "walk through" the shooting schedule point by point and scene by scene, discussing exactly what will physically take place in each case. The beginning of such a walk-through conversation might go something like this:

DIRECTOR: At 6:45 tomorrow, who will be showing up and in what order?

AD: Crew call is 6:45; talent call is 7:30.

DIRECTOR: OK. Let's confirm the crew and positions.

AD: John Martin, DP; Bill Davis, sound; Darla Orren, gaffer; Rudy Gomez, grip; Tom Arton, prompter; and me.

DIRECTOR: OK. And the crew will get busy right away loading the trucks?

AD: Right. The equipment was delivered today.

DIRECTOR: Everything checked out?

AD: Right. It's all in the prop room. The truck reservations are on your desk.

DIRECTOR: Good. And the cast?

AD: Host, Dorothy Miles; installer, Gina McCoy; and customer, Marvin Aganell—all booked and confirmed for 7:30 tomorrow.

DIRECTOR: You're meeting them and showing them to wardrobe?

AD: Yes. And I have coffee and rolls arranged.

DIRECTOR: OK. And how about the client?

AD: She'll be here after lunch. She can't escape a morning meeting.

The conversation would continue in exactly this way until the two had talked their way through every aspect of the shoot. Notes would be made to follow up on any items found needing attention.

Preproduction meetings with the full crew are still another confirmation check. The director should bring in the entire crew half an hour early to go over each scene and any special aspects of the shoot. Again, if the shoot is a simple one, a brief discussion will be adequate. If the setups are complicated, however, flip charts, photocopies of blocking diagrams, or storyboards may be necessary.

For instance, the director might stand next to a flip chart and easel and say:

DIRECTOR: We plan to finish with the host material at noon, so after lunch, Tom, you become a dolly grip. I've got two dolly shots set for the afternoon. One is scene 3. It involves about 20 feet of track, and there are seven extras. Here's generally how it's blocked. The actor moves across this way, and the extras cross back here. It's going to be a tough one. On scene 4, Bill, it's going to be real hard to get a boom mic in. There's a low ceiling, and it's a confined space. Should we think about using a lavaliere?

Discussions like this allow the crew to gain valuable insight into the coming day's activities. They also provide a forum for input that may help the shoot run more smoothly. For instance, the sound person might respond to the director's previous comment about the low ceiling like this:

Wishy-washy Case Study 6

One director I worked with tended to go into his shoots unprepared and undecided about what he wanted and how he should go about getting it.

His typical style was to depend almost entirely on the DP. I overheard several conversations in which the DP would say something like: "OK, what shall we do on this shot?"

"Well," the director would say, "I guess the actress will come in from over here."

"But that will be hard to light. The background is real hot."

"Right. Well, maybe we should have her come from this side."

"Are you sure that will cut OK with your last shot? Wasn't her screen direction just the opposite?"

"Actually, I think you're right. Well, let's see. What do you think—from here?"

It didn't take long for this director to lose the respect of his crew members and his DPs. In addition, his clients listened in on these conversations, as did his producers. The short-term result was a reluctance to work with him by crew members. The long-term result was a lack of work that put a career-damaging cap on his future in the business.

BILL: A lavaliere will probably be fine if it will match your other audio. If there's a direct cut to a scene mic'd with a shotgun, though, I'd say let's try to get the boom in low or go with a lavaliere on the other shots.

Based on this exchange, a plan could be formulated that might save a good deal on time at the location and result in better quality recordings.

The Director as Supervisor

Though his administrative and supervisory skills may overlap, a director tends to become less administrative and more supervisory as preproduction ends and the production itself begins. At this point, the planning and organization are complete and all that remains is execution.

The effective director supervises his crew by setting the pace, establishing a positive tone, and demanding the best from his people. He avoids the temptation to become aloof or prone to tantrums or irrational behavior. The best directors are simply hardworking individuals who inspire their crews by gaining respect. And just how, you might ask, is this done? With plain old hard work, talent, and the use of a few basic supervisory steps.

Communicate

Nothing is more frustrating for crew members than to be left in the dark. Always tell crew members *why* what they are doing is important and *how* it fits into the overall production plan. Besides leading to much more motivated crew members, this will also cut down on rework due to mistakes or misinterpretations.

Just as crew members dislike being left in the dark, a director dislikes returning to the set from a meeting with an actor to find the wrong setup in place. Clear, simple communications will help alleviate these types of frustrating and often costly problems.

Be Accessible

Never create an egotistical or a stand-offish air. Some directors I have met appear to feel this is a must in order to maintain their sense of control or importance. Actually, the opposite is true. If crew members feel uninhibited in approaching a director about a possible change or potential problem, a much more productive shoot will usually result.

If the director decides not to take a crew member's suggestion, it is usually enough to say something like, "That's a good idea, and it would would work if I had something else in mind. But this sequence is going to play out a little differently in the final cut. But thanks. Keep the ideas coming."

Usually such a response is accepted by the crew member, and the productive, open channels of communication remain intact.

Be Decisive

Crew members, especially DPs, will quickly lose respect for a director who is unsure of what he wants or is generally wishy-washy.

If a DP approaches a director on their arrival at a new location and says, "OK, so what's happening here?", she would most likely respect a director who replied:

> We'll do this scene in a single shot. I want to start with a close-up of the calendar here on the desktop. The actor will start talking, and we'll tilt up to a close-up of him. After a few seconds, we'll dolly back and reveal the second actor coming in from back here. I'd like to get a nice smooth master shot we can use for the whole piece, but I may want a little coverage too—maybe an over-the-shoulder shot from back here and a reaction shot.

As the director, you are in charge. Therefore, make sure you know what *you are in charge of. Organize and incorporate all of those* whats *into your shooting script.*

—Donna Chandler
Director

That same DP wouldn't have much respect for a director who responded to her question with an answer like this:

> Well, we need to shoot the two actors here at the desk. I guess one could be sitting. What do you think—would that work out? I'm not sure if they should be sitting or walking, actually. Maybe he could walk and you could follow him—how would that be?

This second director not only has not taken the time to think out and plan his shoot but also seems to have no problem placing his own responsibility for blocking the scene squarely on the shoulders of the DP.

Be Reasonable

A director who wants a $10-million feature on *$10,000* budget is just as bad as one who has no idea of what he wants.

A director who knows his craft knows too what a crew can and cannot accomplish in a reasonable amount of time. While he may challenge them if he himself is restricted by a tight budget or severe time constraints, he will also do everything he can to plan his shots so that they can be accomplished successfully with the resources at hand.

Any DP, grip, or gaffer would become quickly demoralized after a few comments like:

> OK, we've only got a half an hour here, but I need a hand-held shot where they walk through the door, down the hallway, and into the rest room. We'll follow along from the rear and then get another angle from the front. Oh, and we need to shoot an overhead shot in the bathroom from a ladder, and I want a few cut aways. We also need to light this really carefully, you guys. I want it to look nice.

Recognize Good Work

Recognition is a surefire motivator in just about any situation. When people go above and beyond their job description or do excellent work, let them know you appreciate it—and by all means, mention it in front of the rest of the crew.

If the director is respected, his crew members look up to him. A pat on the back and a mention of a job well done from someone we respect can be worth an entire day's paycheck. By contrast, the director who never has a nice word to say will soon discourage his people from putting out that extra effort that can often bring a shoot in on time or make the difference in getting a particularly complicated or difficult shot.

Have Fun

Corporate television production can be a highly rewarding profession. It can also be a tough one, due to constant pressures, frustrations, and numerous other difficulties. Whatever else it is, it should be *fun*. This means that no matter how far behind your shoot may be or how hard a scene may be to get, it never hurts to take five and crack a joke or make light of the situation.

Just relieving the pressure in this way can often revitalize a cast, crew, and the director in a difficult production situation. Aside from all these reasons, *no* profession is worth a heart attack or ulcers.

Say Thanks

I make it a point to thank my crew members after each shoot. Again, a simple word of acknowledgment lets the grip or gaffer who worked particularly hard know that you saw what she did and appreciated it.

You will probably want to book the same people again at some time, and, of course, you hope they will work hard for you again. A parting thank-you can help ensure that this will be the case.

> *Communicate clearly with cast and crew. After all, it takes many people to turn the big picture in your head into a finished program.*
>
> —Donna Chandler
> Director

The Director as Coach

The director *supervises* his crew, but he *coaches* his talent. While these two activities have similar means and ends, the differing factor springs from the word *performance*.

> *Strive to keep your talent happy. They are the most direct link to your audience.*
>
> —Ralph Phillips
> Writer/Director

Crew members work on an ongoing basis throughout the day. Their labors are certainly important and by no means easy, but those labors are also spread out over many hours. Crew members might be thought of as operating on a labor plateau that rises to one level in the morning, stays there throughout the day (with the exception of break periods), and drops off as the shooting day comes to a close.

An actor, on the other hand, is only "on" for a short but extremely intense and critical period of time—much like an athlete. Like an athlete, when her performance is over, she retreats into a resting or replenishing mode, waiting for the next call for her brief, crucial delivery. This means that her day is more like a series of radical peaks and valleys. She is either operating at her pinnacle or in a very sedentary state.

In an athletic sense, coaching means encouraging and inspiring a team or individual to deliver an excellent athletic performance for a *short but sustained* period of time. In an acting sense, remove the word *athlete* from this statement, and the same definition holds true.

The coaching techniques the director uses are also similar to many used by athletic coaches. She *communicates, encourages, demands excellence, recognizes worthy efforts,* and *knows when her actor has reached her limit.* In doing these things, she also *gains her actors' respect.*

As mentioned in Chapter 6, *Human Aesthetics,* when dealing with employees as talent, the director uses a somewhat different tact. We will discuss this aspect of her work in greater detail later in this chapter. For the moment, however, let's concentrate on professional talent.

Communicating with Actors

Communicating means making your ideas clear to another person. This is especially critical for the director, since she is communicating what her actors must then manifest in themselves before the camera. Communicating can also be especially difficult for the director, since she is often explaining or exploring intangible elements, such as emotions and personal opinions.

A director overcomes these difficulties in two ways: She learns to *articulate her own feelings with clarity,* and she learns to *speak the actor's language.*

Articulating To cite an example of clear articulation, I remember a voice recording session I once directed as a newcomer to the field of corporate video production. The actress I was working with was a very talented woman, but she was not providing the delivery I was after. I had a clear idea in my mind of how I felt the delivery should change, but I had no idea of how to explain to her what I wanted.

The sound engineer I was working with was an old pro, and he saw my frustration. He listened as my conversation with the actress went something like this:

ME: It's not quite right. I mean, it sounds good, but it's, well, I'm not sure.

ACTRESS: Am I missing some inflection or tone you're after?

ME: I'm not sure. I think the tone is fine. There's nothing specific, but it's too, well, too abrupt or something.

ACTRESS: Too abrupt. . . . OK, let's see . . .

This went on for some time, and though I was stuttering and stammering, the engineer seemed to sense what I was after. Finally, as I was saying, "Try to be . . . a little more . . . ah . . ."

"Intimate," he said.

As soon as he said the word, both the actress and I knew he had managed to articulate what I had been trying to say. *Intimate* was exactly right. The actress was then able to deliver what I wanted, and the session moved along fine from that point.

How do you learn to articulate such subtleties as this? One way is simply by observation and practice. Think about the feelings and emotions you ask your actors to evoke, and search for the precise word choices that would describe those emotions to you.

Remember too that the description of how a line should be delivered need not always match the visual aspect of the program. For example, in the situation just mentioned, the pictures being shown as the actress delivered her lines were not intimate at all. They were actually visuals of a small town in the days following a destructive earthquake. The intimacy in the actress's voice, however, registered very appropriately when heard with the pictures.

Speaking the actor's language Speaking an actor's language often involves having an awareness and basic knowledge of the type of training the actor may have had. Most professional actors have been trained in variations of two primary schools of discipline: They are either *method actors* or *character actors.*

Method acting refers to the Stanislavsky method developed originally by Konstantin Stanislavsky, a Russian actor, director, and teacher. In the late 1930s and early 1940s, method acting took root and grew in America thanks to a small group of revolutionary actors and teachers—most notably, Sanford Misner, Lee Strasberg, and Stella Adler.

Method acting requires that actors vividly recall events, people, and moments in their lives through sensory exercises. When getting "into the moment," an actor meditates on details of this past event. He then brings those true original emotions into the present and applies them to the scene being acted out.

For instance, an actor doing a scene in which he must display sheer joy on receiving a gift might recall the actual moment in his life when his mother presented him with a new puppy. In order to convey those emotions with true credibility, the actor would get "into" that moment by recalling the type of day it was, the look of the room he was in, the smell of his father's pipe, how the puppy first licked his hand, what its fur felt and looked like, and so on.

Method actors usually prefer a more relaxed and quiet environment on the set. They also like the time to be able to meditate on past experiences, even including smells, touches, rooms sizes and characteristics, and the like.

Character acting, on the other hand, means essentially developing a character profile based on the situation called for in the scene. Rather then being offered a quiet place to get "into the moment," a character actor might require a detailed description of the character, along with an account of his immediate and perhaps distant past history—where he was born, how he was raised, what his attributes, faults, drives, and fears are.

Allowing flexibility Whether it is method or character acting, the director should, first and foremost, be able to talk the actor's language and, second, give the actor as much flexibility as possible in creating the performance. This allows the actor to use his creative talents in developing a performance he will be truly proud to present.

Giving an actor flexibility means not demonstrating how to deliver a line or express an emotion but rather giving the actor certain guidelines and letting him develop the detailed aspects of the performance himself. For instance, it is preferable to *not* say, "Do the line like this," and then read it the way you want it. This limits the actor to mimicking rather than creating true character traits and emotions.

To avoid this, the director might offer a subtext or underlying meaning to the scene, such as "John [the actor playing opposite] wants your recognition. I think when you deliver this line, 'Well, you finally made it,' you're really *congratulating* him for all the good he's done." Asking the actor to use a congratulatory subtext allows him to develop his own means of delivering this line in accordance with the director's wishes.

The use of subtext to alter or "play" with a scene can also be a rewarding experience for both the actor and the director. At times, when a performance just doesn't seem quite right for some reason, applying a new, sometimes totally different subtext can create an unexpected and extremely interesting performance. For instance, if a scene involving betrayal by a friend is being played with a "destructive" subtext, what might happen if the director asked that the same lines be played as a joke? or even begging forgiveness? You can imagine how some interesting variations on the same basic performance would result.

Motivation *Motivation* is another common word in the actor's and director's vocabulary. It refers to the purpose for a line or action. For instance, in blocking a scene, the director may say, "I think you should turn away from him at that point and start for the door."

The actor may then ask, "What's my motivation?"

The director would probably respond with something like, "You've been wanting to get out of the house all through this argument, haven't you? This is the first moment you could actually seize to go. Do you think you'd grab it?"

Motivation should always be in keeping with the character makeup of that person. For instance, in the example just mentioned, an argumentative type of character might not want to leave the room. He might be motivated to stay and argue. A timid or less verbose character, on the other hand, might want to do what the director has suggested.

This interplay of motivation and character is often the subject of repeated discussions between the director and his actors as they develop and execute performances.

The "real world" As much as a director may want to follow the accepted creative etiquette, corporate television time constraints or the production situation may limit her ability and sometimes willingness to do this. If she is behind schedule and losing sunlight and she needs only a single line from an actress, it may simply be easier and quicker to just say, "If you don't mind, I'd like to give you the reading for this line so that we can move on," rather than going into a long dissertation about character, motivation, subtext, and so on.

Actors who work in corporate television usually have no problem with this, since they understand the types of pressures the director is often under. In this case, if the line were "Andrew is really growing up," the director could then say something like, "Put more emphasis on 'really' and less on 'up.' I'd prefer that the line read, 'Andrew is *really* growing up,' instead of 'Andrew is really growing *up*.' "

Whatever the case, try to give your actors as much time and creative flexibility as the situation will allow. Not only does this permit them to truly practice their craft for you; it also shows your confidence in them. Both these elements are positive motivators.

Encouraging Actors

Encouraging actors usually means either spurring them on to deliver exceptionally difficult performances or encouraging them to take a chance by going further with an already adequate performance.

Encouraging an actor can often be accomplished by taking 5 minutes away from the set to discuss ways of overcoming some difficult aspect of the performance. Together, the director and actor might focus on the difficulty and find a way the director can help. Maybe it means shooting the scene in two or three shots instead of one long take. Or maybe it means changing a line or two in the script. It may also mean eliminating a line or movement the actor feels is unnatural.

I once had an on-camera host who had a very difficult time saying the word *terminal* in a line he was supposed to deliver to the camera. Although it may sound absurd, we spent a good 15 minutes and went through 11 takes trying to get the word right. In every case, however, he would pronounce it "ternimal."

Since the program dealt with telephone terminals, I was hesitant to consider changing the line. As the takes mounted, however, I could see that the frustration was wearing on the actor. Finally, I asked that the crew take five, and the actor and I went over the line several times carefully. It turned out it was not the word *terminal* itself but the alliteration of the "T" sounds in the line: "Trying out every telephone terminal in the area could be the answer."

We found that by changing the word *telephone* to *phone*, he had much less trouble with it. I made a decision to change the script and told the actor I was perfectly confident he could do it. He read the line with no problem, and we were on our way.

In another situation, with a few brief comments I once encouraged a woman who was doing a dramatic role for me to go beyond what she felt she was capable of. Although her performance was working, I felt she could do more with it. After the fifth take, I said, "That's a buy. Very nice. Now I want an alternate—a totally loose read, just for the hell of it. I want you to really let go on this one, Meredith. No holds barred. We've already got the keeper on tape, so you're home free. Just have some fun with it."

We did the scene again, and her performance was better but still not there. "You're going halfway," I said. "Pull out the stops! Don't worry. If it's overbaked we just won't use it!"

She did it again, and this time it was brilliant. Her emotional range and all the animation and inflections that came with it blossomed into a truly rich and credible performance.

Finally, a director can encourage his actors simply by displaying confidence in them. If a performance is not working, a comment like "I know you and I know what you're capable of. You can do this scene perfectly" can go a long way toward bolstering an actor's ego at a particularly delicate time. You might then go on to say, "Let's just take a minute to sit down and figure out what's holding you back."

Demanding Excellence

An actor often knows when a director is accepting a performance she is not totally happy with. Unless the conditions are extremely difficult, the actor will usually lose some respect for a director who does this.

The reasoning is that the director should be prepared to work with his actors to achieve solid performances, and not give up too easily if they don't happen right away. To do so is to "rob" the actor, in a sense.

A director can encourage excellent performances in several ways. I often have short, tone-setting meetings with my casts before a shoot begins. In them, I say something like:

> We have an extremely talented crew and cast. I know everyone here is capable of excellence, and that's what I'm going to be after. We have the time and the resources to create a truly excellent program, but we'll all need to put out the extra effort to achieve it. If something isn't working and we start getting into high take numbers, just try to bear with me. I think if we hang in there until its right, we'll all be much more proud of the outcome.

Another way to demand excellence is simply to be direct with an actor on the set and perhaps say something like, "That was a fair performance. In another situation, it might do. But I know you can take it a step further, and that's what I want. I don't want your good or your fair; I want your *best.*"

Recognizing Effort

When the actor does put forth that extra something that results in a wonderful performance, don't hesitate to let her know how pleased you are. Remember too, however, that some actors can sometimes be extremely sensitive. If one sees you showering accolades over another, he may tend to get jealous or resentful. This shouldn't restrict you from recognizing a worthy effort; it just means you should remain aware of the temperaments of the actors on your set and spread the recognition around.

I once worked with an actress who delivered very consistent, excellent performances. She was so good that as the shoot went on for several days, I found myself forgetting to give her much attention. She was so rock-steady and credible at what she did that I began to take her for granted. During that same period of time, I was congratulating several other actors for delivering performances that in the final analysis weren't even up to this woman's level.

On the fourth day of the shoot, I noticed that she seemed a bit more reserved than normal, although her performances remained excellent. Realizing that her quiet mood might be the result of my lack of attention, I called her aside during a break and said, "I just want you to know you're providing the foundation and the example for this entire cast. I realize I'm not saying much to you, but that's because I'm busy trying to get some other people up to *your* standard. You're really excellent at your craft, and a delight to work with."

She beamed and, I knew at once, understood. Her happy mood returned, and she finished the shoot with even more flawless performances.

Knowing When to Let Up

Regardless of all your efforts and those of the cast and crew, there will be times when you feel it is best to simply accept what you have "in the can" and move on. This is usually the case when it becomes obvious that a performance cannot be improved and that time or other constraints are becoming negative factors. If you simply have no more sunlight, for instance, and you have a performance in the can that is adequate, you may be forced to buy that performance and aim for better casting or perhaps rehearsal on the next production.

Since achieving true flawlessness is a rarity, this process of knowing when to let up and buy what you have is not an exception but rather a continual pressure for any director. We will discuss it in more detail in Chapter 11. For the moment, remember that just as with all other creative activities, there comes a point of diminishing returns. This means that when the extra time and effort you may be about to put in will in all probability *not* pay back enough to make it worthwhile, you must have the presence of mind to simply let go and move on.

Letting up for the moment is another strategy. If your take numbers are getting into the double digits and you sense that your actor is becoming frustrated by his inability to get it right, you might just say, "Let's take five and have a cup of coffee." Often, with the pressure off for a few minutes, the actor can gather his thoughts, brush aside the frustration, and garner new levels of confidence. Then another take or two will often be the one.

Gaining Respect

If there is one thing every coach should have, it is the respect of his athletes; the same is true for every director. If your cast respects you, they will give you the 110% required to achieve excellence. If they don't respect you, that level of effort will be the exception rather than the rule.

How, then, do you as a director achieve respect? Mainly by doing all the things we've been discussing in this chapter—communicating with your cast, supporting and encouraging them, guiding and challenging them toward excellence. In addition, however, you can also know your craft extremely well and give 110% yourself.

Having a good deal of craft knowledge breeds admiration and respect among both crew and cast. If you are truly a professional, they know they can trust in your requests and decisions as the right ones for the good of the project. They can also feel secure that their efforts will be guided properly and used correctly later, in postproduction. In other words, they will feel secure that you will produce a show that will create a positive reflection on their efforts and abilities.

Giving 110% yourself should be self-explanatory. If you're not willing to put in your very best effort, the project isn't worth the time and effort on the part of your cast and crew either.

Coaching Employee Talent

The director's coaching techniques for company employees should be much more attuned to creating a warm, friendly work environment than a professional dialogue. As we have established, employees have no acting skills. Thus, to talk with them in terms of character traits, motivation, sensory perceptions, and so on would be useless as a means of changing their performances.

What do tend to work with employees, however, are three simple techniques: *downplaying the production mystique, creating a relaxed environment,* and the occasional use of a few *simplified acting techniques.*

Downplaying the production mystique In the majority of cases, focusing on the mystique of the production process creates tension for employees. When an employee "actor" steps onto a set or location for the first time and suddenly finds herself among lights, cameras, monitors, flags, nets, cables and so on, the sheer enormity of what she is about to do may leave her virtually speechless. The director who plays up the production mystique usually heightens this sense of awe. The end result is almost always a tense, obviously staged performance—even if the employee is simply performing some demonstrative job function.

What is the proper approach when an employee walks on the set?

Creating a relaxed environment I typically greet employee talent to my sets with a warm smile, a friendly handshake, and a comment like:

> Hi, you must be John. I'm Ray, nice to meet you. Let me introduce you to the crew here.

Following these informal introductions, the initial awe of stepping into a production situation often begins to wear off. I might then follow with something like:

> We've got a few simple things you can help us with, John, but we'll have you out of here in no time. So how's it going? You folks as busy in your department as we've been out here today?

This type of inquiry often starts a brief work-related conversation that helps put the employee even more at ease. Once I see this happening, I then begin to ease into a discussion of the employee's role as talent. I might do this with a statement such as:

> Well, what we've got here is a part where we need someone like you to just step up and greet this customer like you normally would. That's really all it boils down to—doing what you do on the job.

I would then walk the employee through the part step by step and perhaps follow with a remark like:

> I really *don't* want any acting. Honestly, what I hope to get is just exactly what you do, the way you typically do it. Nothing corny or set up.

If I've been placed into an unavoidable situation in which scripted lines are a necessity, I might show them to the employee and say, "Is this the way you'd normally greet a customer?"

If the answer is "No," we would set about changing the lines as needed. If the answer is "Yes," I would follow with:

> Okay, good. So show me how you'd do it. If these lines are close, you can go ahead and say them, but if you'd normally say something else, or add anything, just do that.

Through a process like this we would soon arrive at a situation in which the now *relaxed* employee would, as much as possible, be able to *demonstrate* what he normally does versus trying to act.

If the resulting performance is acceptable the shoot can progress as planned. If the employee begins having difficulty, however, there are a few possible options still available to the director.

Using simplified acting skills In some cases an employee can actually be directed like an actor. I've found this most successful when the employee becomes relaxed and comfortable in front of the camera, but she just can't seem to make the lines work naturally. In such cases, the performance might be improved by using a simplified version of a method acting technique.

As you will recall, the method actor concentrates on past events in her life and applies these sensory perceptions to the scene at hand. Keeping this process in mind, the director might say to the employee who was having difficulty greeting the customer naturally:

> I'll tell you what. Think back to a specific customer you recently greeted like this. As you remember that person, just try to remember what you said and how you felt when you met her. I mean, you were probably feeling positive and relaxed, and maybe

Case Study 7 *A Talent Director*

I knew a director who was excellent at working with actors but not very good at audiovisual aesthetics.

Typically, he would spend a great deal of time and energy on the performances of his cast but be unsure of what he wanted in terms of focal lengths, camera movement, and so on.

He managed to make this arrangement work for two reasons.

First, his casting choices and the performances of his actors never failed to be extremely credible, and since actors' performances are a primary concern to producers, this kept him in demand.

Second, whenever possible he worked with DPs who were used to and thus skilled at this type of directing style. They were the types who did not mind working out shots, focal lengths, screen direction, and so on as a collaborative effort, rather than strictly at the director's command.

Whenever this director worked with a new DP, he was sure to mention in advance that this was the way he preferred working. He would follow this by asking if the person had any aversion to it.

Most times, there were no problems. In cases where there were, he simply chose other people for his next shoots.

---

she had a pleasant expression on her face or a nice voice or something.

If the employee can be gently coaxed into recalling a situation like this, as long as the action is kept simple, there's a good chance the performance can be improved.

I have used a similar technique in coaxing employees to perform better in *interview* situations as well. In one case, the employee was doing a fine job of answering my interview questions in a pleasant and articulate manner, but her voice and presence were extremely flat and monotone.

After several takes, I said:

> Okay, very good. Now I'd like you to do me a favor. Ever had a situation where you're talking to someone like maybe a friend or neighbor, and what you're saying is important, but it just doesn't seem to be sinking in? And you know how you kind of wanted to lean forward and say, "Hey, listen up here! This is important!" Well, as you answer my questions this time, consider me someone like that. I'm just not getting it, and you have to drive the point home a little harder to make me understand. Okay?

Again, this is a simplified technique that, if properly handled, can often boost an otherwise flat interview delivery.

In the final analysis, coaching employee talent into delivering credible performances can be one of the corporate director's greatest challenges. The wise director handles such situations with care and sensitivity. He realizes that employees are not actors and to attempt to fit them into such a mold is most often embarrassing for the employees and detrimental to the production. The wise director also knows, however, that with careful, insightful handling, employees can add a level of credibility to corporate programming that few, if any actors can supply.

Finally . . .

The director's people skills are really what make or break him. All the technology and special effects in the world cannot make a good program out of poor acting and technical mediocrity. But when the director is able to work with people on a challenging, positive, and productive level, he is given the ability to take his program to great levels of excellence both at a technical and a human level. Once he is capable of this, he need only be concerned with one more critical skill—his ability to make sound, creative *judgments* under some of the most difficult circumstances imaginable.

11 Judgment Skills

So far, we have looked closely at the business aspects of corporate directing, the aesthetic qualities the director must concern herself with, and the people skills she must master. In all these cases, we have talked about *quality* performances, *excellent* composition, *perfect* tone, and so on.

But what exactly are the manifestations of such words? Just what separates excellent composition from, say, good or mediocre composition? What elements could be construed as being intrinsic to a quality performance? And how do they differ from those of a bad performance, or just a pretty good one? What is perfect tone as compared to acceptable tone?

In the final analysis, much of what separates these definitions is *subjective judgment*. This means that it is in the eye of the beholder and it is different for each of us, depending on our personal tastes and opinions. It also means that if three, five, or ten people on a production crew watch a scene, all will have different opinions about whether or not it works and, if not, what is wrong with it. This, of course, means that someone must make the final judgment. This final judgment responsibility is the director's, and it may be the single most important artistic skill she brings to a shoot—the ability to know what works.

It is often this one critical element that separates a director from other members of the production crew. She has the ability to watch a scene play out, visualize how it will look when edited, and decide whether it will successfully convey the script interpretation to her intended audience while at the same time keeping her client and producer happy.

Although this judgment skill may be the most important of the director's abilities, because of its subjective nature and the fact that it seems to be a product of experience and personal sensitivity, it is also one of the most difficult to explain in the pages of a book. While we may not be able to define it exactly, on the following pages we can definitely explore some of the sensitivities that seem to be intrinsic qualities in directors who know what works.

Sensitivity to the People and the World Around Us

Any director who must judge an actor's performance must know how people might act in a similar situation. For instance, if a married male employee is being fired from his job, how would the typical wife react when she was told? Would she cry? get angry? become panicky about the future? Would she want to call the boss and yell at him, or go sit in church and pray? Or would she simply sit down and begin to laugh?

Each of these are possible reactions to a situation a corporate director might find is her script. Each reaction is also related to the character development and motivation mentioned earlier.

For instance, is the wife proud of the husband, or has she despised him for some time? Is she a confident woman, or a very negative and skeptical person? What does she think of the company her husband works for? Has she wanted him to quit for a long time anyway, or was this job the very best thing that has happened to him and her in years? You can see that the possibilities are numerous, and the director must be sensitive to how this wife character might react with any one of these character profiles.

Besides human aesthetic judgments, there are also countless audiovisual judgments to be made. Will a close-up have more impact shot in a long focal length or a short one? Is a close-up even needed or appropriate? Will a dolly shot heighten a sense of tension by following the action, or would a zoom or a series of lock-off shots be just as effective? Is hard, dramatic lighting called for, or is flat lighting?

The ability to make these and many other similar judgments is at least partly an intrinsic skill that comes from the continual observations of a person sensitive to the people and the world surrounding her. She either has seen and pondered how persons in real-life situations act or she is sensitive enough to the emotional content of a scene to visualize a simulation of those actions accurately enough for it to become a credible part of her illusion.

Sensitivity to Audience Tastes

The director must also be keenly aware of how her audience feels about what she is directing. Consider the director who has been charged with the responsibility of shooting a short, year-end recap by the CEO on how well her roofing supply company has done and how bright the future looks.

If the audience for this piece is craftsworkers who drive forklifts all day, would the CEO be more effective giving this message in a warehouse situation or from the executive boardroom at the head of a large oak table? The warehouse setting would probably be most effective, since it would bring the executive to the audience's level, placing her in familiar territory. On the other hand, if this same message were being presented to a group of executive board members, the CEO would probably be most effective in an executive setting familiar to them.

Although such location decisions are often made in advance by the scriptwriter and producer, many times they are not. When they are left to the director, she should be sensitive to her audience's likes, dislikes, and opinions about the subject being shot.

Another example in this same area is wardrobe. The director normally decides on her actor's wardrobe. If she is directing a company news program featuring an on-camera host, how should he be dressed? Again, this comes down, at least partly, to a sensitivity to her audience.

Would he appear most credible to them in a three-piece suit or wearing a tie and white shirt with the sleeves rolled up? Or should he perhaps be in a V neck sweater with no tie or a waist jacket? Should he wear semiformal slacks, such as corduroy, or expensive slacks? tennis shoes? or boots?

Each of these is a judgment the corporate director will be called on to make, and each of them can either enhance the illusory power of her program or detract from it.

Sensitivity to the Client's and Producer's Tastes

When the director has satisfied her own sensitivities and those of the viewing audience, the final test is, in a sense, the client and the producer. Failing to take their preferences into account can be as damaging to her program as choices inappropriate for her audience.

For instance, if we return to the CEO year-end wrap-up mentioned earlier and consider it from the client's point of view, the director's first choice might change. This could be the case if the client is a conservative corporate executive, such as the public affairs vice-president. This individual might be very concerned about the CEO's image, and although the director may feel strongly that a warehouse setting would be appropriate, the client may object on the grounds that it's too "rough" of an image for a CEO.

Although the director may totally disagree with this observation, she may have to concede to her client as a matter of basic corporate politics. As long as it is done tactfully, there is no harm in voicing her opinion as to why she feels the "rougher" image is appropriate, but in the end, if the vice-president disagrees, she will have to change her location.

This same type of sensitivity "reading" must be taken on the producer. Is he comfortable with what the director has decided? If not, why, and is there a possibility of a compromise that satisfies both? Just as with the vice-president, if the producer is open to suggestions, there is no harm in presenting reasons why the director feels she has made the right choices. But if the producer is unwilling to budge, the director has little choice but to give in.

Keep in mind that, as we have said several times, giving in to either clients or producers should not be considered a major artistic disaster. A corporate program is an important creative effort, but it is not a work of art worth damaging a career or reputation over.

Further, regardless of the director's intuition, there is always the possibility that the client or producer has some larger perspective that may be valid. For instance, suppose the CEO finds it personally awkward to be seen "acting" in front of employees. The public affairs vice-president may know about this quirk but be unwilling to pass the information on to the director. Or the producer may know that a union disagreement is developing in the warehouse and that bringing the CEO in at this time is not the most appropriate thing to do.

Judgment Skills Case Study 8

A person who had done a great deal of lighting and camera work for me approached me one day asking for advice on how to get started as a director.

Since I had worked with this person quite a lot, I was aware of what I felt was a general shortcoming he should work on. "You need to work on your judgment skills," I said. I then elaborated, telling him that I recalled a number of times during our association when he had made incorrect judgment calls with regard to clients, producers, and production responsibilities in general.

I noticed that he didn't seem to make the connection between client and producer judgment skills and his directing ability. I reiterated by telling him that a corporate director had to show good judgment skills not only with regard to the aesthetics of the scene he was shooting but also with regard to client and producer handling. I went on to say that this not only was important to a potential director but also was an *inherent part* of directing corporate programs—just like composing a shot or directing an actor.

He said he understood, but evidently he did not, because his judgment continued to be faulty.

Not long after our conversation, he was reprimanded by a client for using inappropriate language in a work area, and shortly after that, he ended up in a confrontation with a producer over how a set should be struck and stored.

To my knowledge, no directing work has yet come his way.

These types of situations are not uncommon in corporate television production. They point up the need for the director's ongoing sensitivity to her clients' and producers' needs.

Other Judgment Calls

What other judgments must a director make? Consider these: What crew members will work best together? Which actor should be cast in the role? Is a set best in this situation or a location? Is the CEO presenting his message credibly, or should he be told he sounds nervous? Should a technical expert be invited along on the shoot? Should the client be sent an audition reel or simply be told what choices you've made? Should the producer be told the budget is too small for what she wants? Should three angles of coverage be shot or a single angle? Should the camera be low or high, close or far? Should sound be recorded or not? Should lunch be catered, or should the crew be released and told to be back at the shoot in an hour? Should the day go longer than 10 hours, resulting in considerable overtime costs?

As you can see, the director's judgment calls play a critical role in every aspect of her production. And as mundane or minuscule as some of those judgments may seem, they can be the very factors that make or break an entire production—and sometimes a director's career.

If the CEO's footage is brought back from the warehouse and the public affairs vice-president doesn't like it, chances are it will be scrapped and a new shoot will be arranged. Not only does that cost extra money, create extra pressures, and make the producer look extremely bad; it also makes almost nil the chances that the same director's judgment will be trusted again.

Judgment Under Pressure

Pressure is a familiar word to any director—corporate or otherwise. In a sense, it is also the "kicker" with regard to all the judgment issues we've been discussing. You see, making the many types of decisions we've explored with *no* pressure applied would be difficult enough, but this is rarely the case for the corporate director. In production, there is virtually always pressure, and sometimes it is excruciating.

Visual versus Content

It can sometimes be a great temptation to buy a scene on the strength of its visual quality alone—even if an actor's performance is considerably less than perfect. For example, consider a scene that requires the camera to start in a wide shot, dolly into a two shot as dialogue begins, and dolly in again to a close-up as one actor exits. From a camera operator's standpoint, this can be difficult to pull off. Not only must the camera move to specific marks at a set rate of speed, but the focus must also be adjusted at critical points along the way and frame edge marks must be found and hit by the DP on the fly. On top of this, if the scene is being recorded on location, any number of extraneous elements can ruin an otherwise good try—airplane sounds, the sun passing behind a cloud, or wind rustling into a microphone, to name a few.

It is not unusual for such a scene to require ten or more takes. During this time, the actors will begin to get tired, the crew will become increasingly frustrated, and the director, producer, and client will become more and more tense with each failure.

Now imagine that the camera move is suddenly achieved with absolute perfection. As playback of the scene is viewed, those who are gathered around the monitor resound with a cheer, and the DP lets out a sigh of relief, at the same time shaking hands with his gaffer.

There is one problem, however: The performance does not work for the director. A line was delivered with a weak inflection, there was a momentary stumble in another spot, and one actor looked stiff and unnatural as he moved to his second position. As the director realizes this and looks up from the monitor, he is surrounded by 5 to 15 people who want nothing more than to hear "That's a buy."

The sound recordist says, "Worked fine for sound. Not a rustle!"

The DP declares, "I'll *never* hit it that good again."

The client chimes in with, "Boy, three o'clock . . . it's getting late!"

The director checks his watch. Sure enough, on top of everything else his shoot is an hour behind schedule, and with the onset of afternoon sun, the quantity and quality of the light are changing rapidly. He therefore has a tough choice. He can buy the scene because it's technically perfect and then look for a way to cut around the poor performances. Or he can say, "First positions. We're going for take 16."

This, of course, is only one of countless similar examples. And what should the director do? Contrary to what you may think, it is not always prudent to go for that take 16. The director's decision should be weighed based on a number of factors. How far behind time is the shoot? Can the loss of time be made up the next day? How bad was the performance? How easy will it be to cut around the bad parts in editing? What extra coverage must be shot to achieve this? Again, the choices are too numerous and diverse to list.

Content versus Visual

The same thing can happen in reverse. A performance may suddenly play out absolutely exquisitely after 14 takes, just when the director is sure the actor is nearing exhaustion. At first, the director breathes a sigh of relief. She has it! It was brilliant! She can move on, thank heaven, and maybe still make it in with only an hour of overtime!

But when the shot is played back, the balloon is burst. The shotgun microphone had dipped into the edge of frame at a critical moment.

The director will no doubt hear comments like "It'll be cropped in duping," "They'll never see it," or "Even if it doesn't get cropped, if you didn't know it was there, you'd *never* catch it." Again, the decision must be weighed. Will it really be cropped? The person who said that will not have to sit in the screening room with the client and producer and take responsibility when someone says, "Did I see something for a second there?"

Or is there a way to cut around it? Does the actor have enough left in him to do it one more time? How is the producer feeling? Can a "pickup" of only a portion of the scene be used?

While you're busy weighing out that dilemma, here's another typical corporate situation—the familiar CEO message. Assume that the director is shooting the top man in the company and he is in a great rush. He says, as he enters the studio, "I need to be out of here in 10 minutes."

He sits on his stool in front of a pre-lit flat, and a microphone is pinned to his lapel. He speaks to provide an audio-level check, and the microphone isn't working properly.

The second recordist goes to work, frantically changing the batteries and checking the cable connections. Meanwhile, an entourage of dark-suited executives crowd around the CEO, the producer, and the director. All look extremely important, terribly rushed, and highly irritable to boot. In fact, the CEO himself has a particularly distasteful look on his face.

The mic is fixed in 5 minutes—record time. The director is ready to say, "Roll tape," when the DP realizes he forgot to set his backlight. There is virtually no separation from the background flat. The shot doesn't

When directing a CEO, be sure the executive is given your full attention. Also, you must exhibit a nonabrasive self-confidence in your abilities as a director.

—Dan F. Johnson, M.A.
Producer/Director

Case Study 9 *Emphasis Where It Counts*

A director I know once did a program consisting of three primary elements: a role-play involving two celebrity actors on a fairly elaborate studio set, a number of people-on-the-street interviews, and a series of very simple shots of employees at work.

Although the scenes involving the studio set and the celebrity actors were the most directorially challenging and rewarding, my friend realized that the other elements were equally, or perhaps even *more*, important to the tastes of his viewing audience—a group of company executives.

Based on this good judgment, he shot all three elements of the program with *equal* emphasis and attention to aesthetic detail. He did not, as some other directors might have, fall into the trap of placing the majority of his focus on the "Hollywood" celebrities.

The result was a well-balanced and thus well-received executive program.

---

look good at all. Six of the 10 minutes is gone as the DP scurries up a ladder into the grid to set the backlight.

Five minutes later, every executive in the place is checking his watch. The producer is developing a nervous twitch. The CEO looks downright angry. The director finally says, "Role tape," and . . . the copy on the teleprompter is wrong.

It is fixed in a flash. The director rolls again, and *finally* the CEO is able to present his entire message. There is only one problem. The performance is noticeably staged and nervous. The CEO looks plain old bad. The five executives gathered around the director, however, immediately start spouting phrases like "Great, J.B.," "Very sincere," "Sounded right on to me," and "I'm convinced!"

The producer looks at the director, and he is sweating. The CEO is checking his watch.

If you are this director, what do you do now? Do you go along with everyone else who is scared stiff, even though you *know* this man will be laughed at later when viewed by employees? Or do you risk both his and the producer's wrath by saying, "I'm sorry, but that one was a bit rough around the edges. If you don't mind, Mr. Billings, I'd like to do it again." Your answer may seem easy now, but, take my word for it, it will *not* be easy when you are sitting in the director's chair.

Summary

These types of "decisions made under fire" are a constant part of the director's life. If she is a sensitive and observant person, her judgment skills will most likely be sound and she will find ways to come away from each shoot with human and technical aesthetic qualities that produce credible scenes and programs.

By contrast, if she is insensitive to her world and the people in it, she will have a difficult time attuning herself to viewer's, client's, and producer's tastes. In this case, her judgment skills will almost certainly be faulty, and she will run into continual problems as the footage she directs is subject to producer and client scrutiny.

So important is this issue that, in the latter case, it's highly likely that in short order she will find herself in a different job market.

PART III

A Corporate Directing Project

The Call 12

Reading about directorial skills offers one valuable perspective on the various aspects of the discipline. Another, equally valuable perspective, however, can be gained from seeing those skills put to work in an on-the-job situation. In this section, we will gain that second perspective by following a complete production scenario from beginning to end.

As in the other parts of this book, in this part we could never hope to cover every aspect or occurrence an actual directing project would entail. But we *can* get a solid feel for how such projects are carried out every day in the corporate world. As you read the following pages, think back to the various skills we have discussed and note how they are applied in the situations that will now unfold.

The call comes at 8:00 A.M. on a Thursday.

John Mitchell, an experienced free-lance corporate director, is watching the morning news, sipping a cup of coffee, and preparing to leaf through a recent publication that lists corporate production facilities.

Since completing his last project, John has been out of work for 9 days, and the "feeler" calls he has recently made are not resulting in work. Like all free-lancers, he is beginning to get anxious.

"Hello," John says.

"Hi, John, this is Angela Sparks from D&L, Inc."

John immediately gets a very positive feeling about this call. He is fairly sure it will mean work. He has done directorial work for Angela before and enjoyed it. The last project was a simple voice-over program, but, as John recalls, it turned out well and was well received by the client and viewing audience. "Hi, Angela," John says. "Nice to hear from you. How are things?"

"Busy," Angela replies. "Which leads me to the reason for this call. Would you be interested in taking on a directing project?"

"Sure. I enjoyed the last one."

"Good. I was pleased with how it came out, and I enjoyed working with you too."

John attempts to gain a little initial information on the project. "What's this one about?"

"It's a little more involved. It's basically a role-play piece that deals with what our governmental affairs department does. I think it will involve some studio and location work—probably about 3 or 4 days. I've got it scheduled to shoot in 2 weeks."

"Sounds challenging," John says. "When do we start?"

"Can you get in tomorrow morning for a rundown on it?"

"Sure. What time?"

"9:30?"

"I'll be there. Will the client be at the meeting too?"

"No. He's got another meeting, but we can get going without him. The script is approved and ready to go."

"Great."

"See you in the morning."

John hangs up and takes a sip of coffee. His concern over when and where the next check will come from has been put to rest—for the moment, at least.

Meanwhile, Angela shares John's positive feelings as she hangs up the phone. She has found John to be an experienced and thorough corporate director on the few projects they have worked on together. In addition, the clients like John. He has a knack for quickly making people comfortable and gaining their trust. These traits, Angela has found, make her job as a producer much easier and more rewarding.

On top of this, the script development phase of this project has gone very smoothly. The writer managed to create an interesting and entertaining scenario to convey the proper information, and the client liked the script very much (the complete script can be found in Appendix 3).

Angela opens the project folder, and on top of the script and other paperwork is a budget (see Appendix 4). She has created this herself, based on a breakdown developed just as the script was nearing final approval.

She notes that the director's fee she has arrived at for the project will cost her just over $6000. This is based on a nonloaded rate of $500 per day for 5 days of preproduction ("prep"), 3 days of location shooting, 1 day of studio shooting, and 1 day of postproduction ("post"). This calculates to a flat fee for John of $5000. The $1035 difference between the two figures (Angela's cost and John's fee) is the load rate, or the cost of payrolling and insurance that Angela must bear as part of her overhead.

All in all, Angela is pleased with the way the project is going. She's looking forward to meeting with John the next morning.

The Initial Producer Meeting

13

The next morning, John is up early and has a light breakfast. Following this, he turns on his VCR and views a copy of the last two projects he did for D&L, Inc. Both were fairly simple as corporate projects go, but they also turned out to be very effective.

One involved an on-camera host who gave a tour of the new company supply facility. The other was a discussion of general safety guidelines with a voice-over narration. This program included several employee interviews and was mostly made with stock footage and slides transferred to tape.

John watches these to refresh his memory in case they come up in the meeting. In addition, they trigger recollections of how D&L, Inc. and Angela like to work. Following this, it's a shower, a shave, and a quick shoe polish. John dresses in a sport coat, a tie, and dress slacks. He checks his briefcase: tape recorder, business cards, folder, paper, pens, and pencils. When everything is set, he heads for D&L, Inc.

As always at D&L, Inc., John signs in with the receptionist at the front door and is asked to have a seat. The receptionist calls for Angela, and several minutes later she appears. "Hi, John," she says with a friendly smile.

John returns the greeting. "Hi. Well, how's everything been for the past—what's it been, about 3 months?"

"I think that's right. June was when we finished that voice-over project, I believe. And it's been busy—*real* busy. How about a cup of coffee?"

"I'd love one," John says.

Five minutes later, John and Angela are seated in Angela's office. John has just been given a copy of the script he is to direct. An initial glance through the pages tells him several things immediately: It's a role-play involving two principal actors; there will probably also be company employees used for minor roles; there are a couple of montages that appear to be fairly involved; and, as is very common, there is an executive "talking head" close.

He has also been given a copy of the needs assessment done by Angela at the outset of the project (a complete copy is shown in Appendix 1). This provides him with a good deal of the basic facts, such as objectives and audience characteristics, that are the foundation on which the script was designed.

As the two sip coffee and John continues to leaf through the script and needs assessment, Angela gives him an overview. "As you can see in the assessment there," she says, "the program is meant to wake employees up to the fact that we have this governmental affairs department and that the people in it work very hard. The client has found that the general feeling among employees is that his reps sit around all day in the 'ivory tower' and spend their evenings eating lobster thermidor at political dinners."

"Pretty cushy, huh?" John says.

"The truth is, it's *not* a cushy job—not by a long shot. That's what the writer was trying to convey, and I think she's done a nice job."

"Looks like the word *respect* is part of this whole thing too," John says. "As I glance through these pages, it seems to me that the writer is trying to create a general respect for these reps. I mean, she's got them working late, hanging in under pressure, dedicated . . ."

"Right on," Angela says, pleased with John's acute sense of perception. "And if we do nothing more than create that sense of respect, the program will be a huge success and you'll have made a believer out of another client."

Hearing this, John turns to the "Objectives" section of the needs assessment. "How about these *instructional* objectives—is the client pretty keen on them too?" he asks.

"The way I read him is this," Angela says. "He feels that if he's going to spend the money for a video program, it should *teach* people something about his department, but his most pressing need is to see his people get respect."

"And how about the audience?"

"I think they need both. That's why the objectives are in there. They need to respect his people, but in order to do that, they need to know something about what goes on in the department."

"Makes sense," John says, thinking to himself that Angela is a good producer who has a keen sense of not only what her client wants but also what's *good* for him. "Tell me more about the client," John says.

"He's a high-level guy, a state operations manager. He's also a very nice person—dedicated and really behind his people. And he's very willing to keep his hands out of the details and let us do our own thing."

"Great."

"We just have to be sure that what we deliver is top-notch."

"Good. I'm looking forward to the challenge."

"I've budgeted your fee at a flat $5000 on this one," Angela continues. "I'm figuring on 5 prep days, 3 location days, and 1 day in the studio. And I'm hoping you can spend a little time with the off-line editor to get him going, and then drop back a few times to be sure the cut is right. For post, I threw in another day at $500.

"Sounds fine," John says. "in the usual thirds?"

"Right," Angela replies. "A third today, a third after production, and a third when the rough cut is approved."

"So how about crew?"

"We haven't booked anyone yet. You know the regular people we usually work with. You'll have an AD, Jenny Martin, starting Monday, so I figured you could let her line them up."

"And the studio and remote equipment?"

"I've tentatively scheduled both based on my breakdown. You'll have to let me know if we need to adjust any days or times so I can firm things up."

John is pleased. As usual in his dealings with Angela, things seem to be organized and well thought out. "Well," he says, "everything seems to be in order. And from what little I've seen of the script, it looks like a nice project. I really appreciate your thinking of me for it."

Angela smiles. She says, "I think of the people who make life the easiest and most rewarding for me, John, and as a director you do that."

"Glad to hear it." John says. "Well, I think I'll head home with the script and get a jump on things."

He and Angela shake hands, and John leaves. It is now 11:20 A.M. John gets on the road, finds a local fast-food restaurant, and, with a hamburger to munch on, spends his lunch in the first phases of script analysis.

14 Initial Prep

On Monday morning, John walks into D&L, Inc. at 7:30. His AD, Jenny, is not in yet, so John gets a cup of coffee from the cafeteria, sits down with the departmental talent books, and begins to browse.

The D&L, Inc. talent books are a group of three-ring binders in which are filed 8 × 10-inch professional pictures known as head shots (a single picture of the actor's face) and composites (several smaller pictures on the flip side of the head shot), as shown in Figure 14.1. These pictures are filed when an actor is cast in a D&L, Inc. program. Along with each picture is a listing of the programs in which the actor appeared. This, then, acts as one source of talent for directors like John when casting new shows.

In reading the script over the weekend, John has found that there are three principal roles: Tom, Phylis, and the governmental affairs director. He has also found a number of smaller or "bit" parts, such as the facilitator in the opening scene, a few additional governmental affairs reps, and the construction superintendent.

As he reread the script a number of times, he began to get a "feel" for how these characters should look and act. With the talent books in front of him now, he begins to try to match the pictures with those initial feelings.

The Tom character, for instance, is basically an unhappy person looking for an out. John feels that since he is in construction, he could logically appear fairly thin or hardened. A chubby or soft look probably wouldn't work for him. Besides being unhappy, Tom seems somewhat lazy and cocky and also closed-minded in that he can't quite seem to grasp the idea Phylis keeps trying to make clear—that governmental affairs is hard work. This attitude and lack of openness on Tom's part could seem natural for an immature person, which means he could also be fairly young.

Phylis, on the other hand, seems to be very mature and businesslike. She has a clear idea of what her job entails; is challenged, it seems, by the hard work it involves; and is quick to spot Tom's attitude problem. John thinks to himself that to make her too attractive would appear somewhat canned and work against a true image of her character. Initially, he sees her as middle-aged, average-looking, and very businesslike in her dress and manner.

The governmental affairs director has begun to take shape in John's mind as a kind of general: a sophisticated, gray-haired businessman, pos-

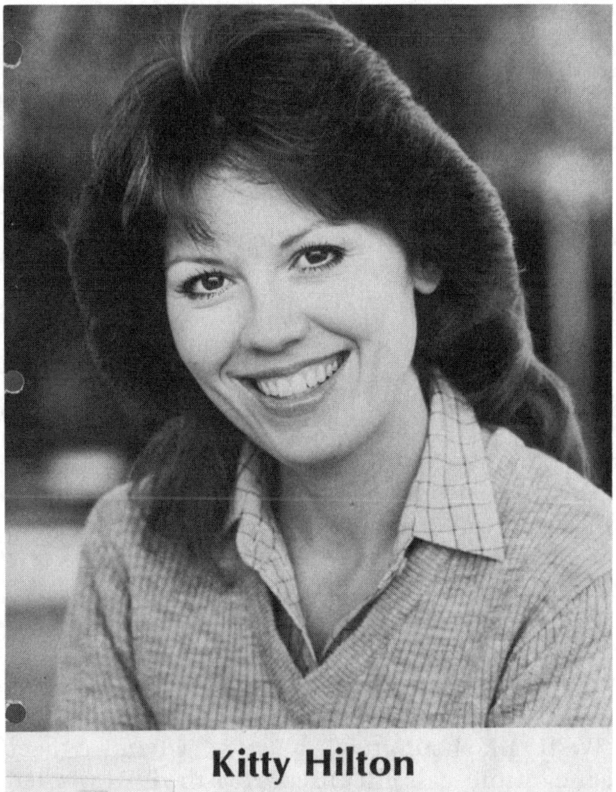

A

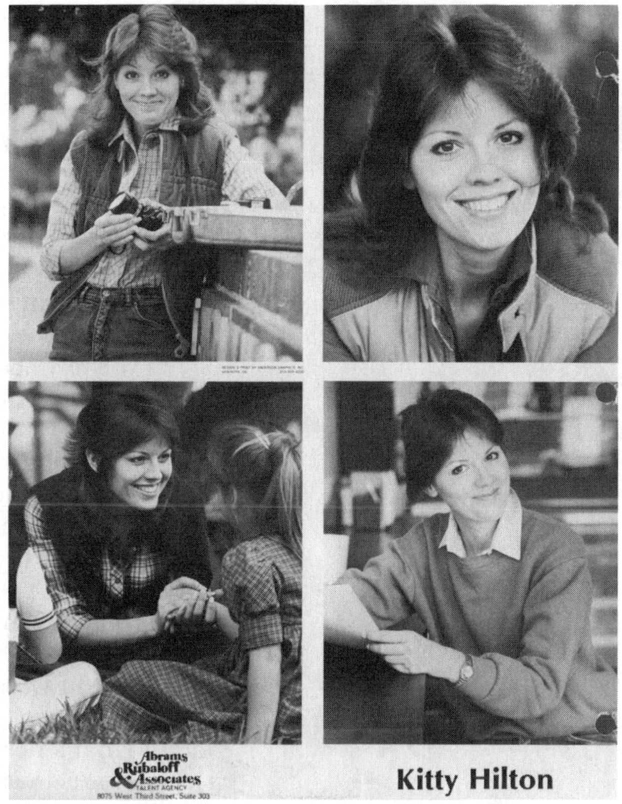

B

Figure 14.1 Head shot (A) and composite (B). Professional actors distribute these to producers and directors, either through auditions, friends, or their agents.

sibly with rough external features and strong sense of command. Under his hardened exterior, however, is a deep sense of dedication to the people who work for him.

With these criteria in mind, John has begun to find some possible candidates among the pictures. As he is pulling them from the book and placing them in a folder, Angela and Jenny approach, and Angela makes introductions.

After Angela leaves, John tells Jenny that he is pleased to be working with her. He then asks that she first acquire a copy of the script and read it over. He tells her that he will be about 20 minutes longer with the talent books and then would like to meet with her.

Half an hour later, John and Jenny are seated in a conference room. As his preproduction assistant for the next 5 days, Jenny will become John's right-hand person.

John first gives Jenny an overview of the project as he sees it at this point. He tells her *why* they are producing the project, *what* the tentative schedule will be, who the *client* is, and how to reach him. They then begin talking over the upcoming weeks' activities.

Being a seasoned corporate director, John knows that his main concerns during this preproduction period should be *casting;* finding, analyzing, and deciding on all *locations;* meeting and working with the *client;*

and making sure that he fully understands and prepares the *script*, ensuring that it has no flaws that will develop into problems during production or postproduction.

In order to assure himself that he will be able to give ample time to these important areas, John assigns Jenny a number of other tasks. "You can start with a script breakdown by talent, locations, props, and anything else unusual you find," John says.

"Fine," Jenny replies. "And how about casting—do you want to audition?"

"Yes," John says. "Let's set that up for Thursday."

"Any idea how many actors?"

"Probably about one person every 10 minutes, all day. I think Angela said we can use the studio."

"Do you want to record them?" Jenny asks.

"Yes," John decides. "But there's no need to use a studio camera and bother an engineer. Let's just use one of the small consumer cameras. We'll need two chairs for actors and a table with a few chairs for you, me, and maybe the client and Angela. I'll get us a mic and put up a few lights. Oh, and you can pull the tapes these people are in for me."

He hands Jenny several tapes, the talent pictures he has collected, and concludes, "Let's meet again in a few hours and talk."

"Do you want me to wait on the crew until later too?"

John thinks to himself that Jenny is a good AD. She seems to know exactly what's needed. "We'll talk about crew," he says, "when we meet in a few hours. But in the meantime, call Bill Olsen. He's the DP I worked with last time. He's got a real good eye, and we worked together well. See if he's available for the production days."

Since this program is almost entirely a role-play, John knows he will be spending a good deal of time with the actors. When the performances are right, he knows he will have depended on Olsen to have achieved the look and the shot. In this way, the two elements—photography (often called videography in video) and performance—will work together as they should when the scenes are actually recorded.

Having assigned Jenny her first activities, John now settles down in a conference room to read the script over in more detail and make notes in the columns. As he looks through the script, John finds that it definitely has a modulating pace. In general, the conversation segments between Tom and Phylis are relaxed, but they also contain an underlying sense of tension. The montage sequences, on the other hand, are written to show how intense and busy the governmental affairs department can get. Those sequences should be much faster paced in both shooting style—perhaps with a hand-held camera—and editing. He notes a light spot on page 7 when a rep requests no anchovies on his pizza; he also notes a revelation on page 10 when Tom makes a call to the superintendent and we find out that his was one of the jobs probably saved by Phylis's work.

Continuing his review of the script, John also finds that the dialogue is in character and naturally scripted. It is kept moving back and forth and sounds very conversational. Tom's character is written a little "looser," as it should be, and Phylis's is scripted as slightly more formal and mature or educated.

Motivation too seems to be believable for each character. For instance, John finds it believable that Tom would approach Phylis at the break and ask her about her job. When the two take a seat a few moments later in the coffee shop, John imagines Tom turning his chair around and sitting on it with the backrest forward. He writes in another brief note to remind himself of this thought later, when he gets into actually blocking the script.

Finally, he notes the executive's closing remarks and makes a note for Jenny to call the secretary and check the CEO's calendar for a good time to shoot him.

John also begins to think in terms of locations as he reads the script. Although he knows Jenny will provide him with a total list of these as a part of her script breakdown, John figures now that there will generally be five locations.

The break room scenes between Phylis and Tom will most likely be shot on a simple set in the studio. John quickly reasons that if what is revealed to the camera appears to be a *corner* of the coffee shop, a coke machine, several small tables, and a condiments bar would work well. He knows of a prop house where such furnishings can be rented on a weekly basis for a reasonable fee. He makes a note to have Jenny check on this prop house.

The majority of the montage shots, John thinks, could probably be recorded in the governmental affairs office, which Angela mentioned is in another wing of the headquarters building John is now in. This thought is attractive to John for two reasons. First, it is close by in the same building, which will make for a much faster and more convenient production. Second, it may allow the department and its employees to become involved in the shoot.

The construction superintendent will have to be shot on a construction site, John decides, or on a mock site that looks authentic. He notes that there are several other office-type shots; the executive close will probably also be done in the studio, John reasons, and the opening scene could be shot in the departmental conference room.

After 2 hours of this kind of script review, John's copy is cluttered with marks but he feels comfortable that what the script attempts to accomplish can be shot convincingly. With these factors in mind, John makes a quick list:

- See Angela—
  - CEO,
  - props rental
- Call client—
  - meet
  - scout governmental affairs department
- Prop house—
  - call
- Rough schedule

SCRIPT BREAKDOWN V-225A GOVERNMENTAL AFFAIRS

TALENT	PROPS
Facilitator	coffee cups
Tom Martin	coke machine
Phylis O'Connel	condiments
Governmental Affairs Director	easement bill (or cover)
Sacramento Rep	position paper (or cover)
Rep #1	typewriter or PC
Rep #2	coke can
Superintendent	pizza
CEO	memo
4-5 extras	

SETS	NOTES
Conference Room	*Check continuity for multiple wardrobe changes for Phylis
Break Room	
CEO Studio Set	

LOCATIONS

1. INT Governmental Affairs Office
2. Legal Department
3. Construction Site
4. Hallway (sc. 13)

CLIENT: Darren Whitesall (213) 555-3712

PRODUCER: Angela Sparks (818) 555-7884

Figure 14.2 Script breakdown.

As he is finishing his list, Jenny approaches with a complete script breakdown (Figure 14.2) and several notes. Looking over what she has brought, John is pleased. "Let's book Bill," he says first. "Tell him I'll call him later to talk about what I need."

"Sure."

"I'm going to meet with Angela, call the client, and do a little scouting. Oh, and call Anderson Rents on La Cienega. Find out the weekly rate on a coke or coffee machine, a condiments stand, three small round tables with chairs, and a few full plants. I'll be back in a few hours."

Jenny has been taking notes. "OK," she says. "By the way, I have a small camcorder arranged for the auditions, and the tables and chairs are in the room with flats and props. Martin, the engineer, says we need a special connection to run a regular mic into the camcorder, but he can give it to us or we can just run off the camera mic."

"The camera mic will be fine," John says. "I think script pages 2 and 3 would be good to use in the auditions. They both have some good characterization and some places where I can ask for variations." With that, John's off toward Angela's desk.

Once there, he vcrifies several things: that it is OK to call the CEO to be interviewed per departmental protocol, that he may check on some coffee shop props for the set, and that he can call and meet the client and scout his department. Angela gives him an affirmative, and he heads for a phone to call the client.

15 Client Introductions

"Hello—Darren Whitesall here."

"Hello, Darren," John says, "This is John Mitchell. Angela Sparks has brought me on board to help with your Governmental Affairs video."

"Oh, right. You must be the contract director."

"Yes," John says. He had purposely *not* referred to himself as the director in his first introduction. This is because he has learned over the years that downplaying the "Hollywood" image is usually a good idea with corporate clients. What they usually want, John has found, is a sense of common ground—a kind of comfort zone. Most times, this is best provided with a mature, businesslike attitude instead of a "glitzy" image.

"So are we on track?" Darren asks.

"Actually, yes, we are," John assures him. "I'm very comfortable that we have an effective script, and as usual, Angela has things well organized."

"I thought so. That's good to hear. So how can I help you?"

"One reason for my call is just to introduce myself and the assistant who will be helping me on the project."

"I see. And who's that?"

"Her name is Jenny Martin. She'll probably be calling over the next week or so about things like props and locations, or maybe even to get some help in lining up technical advice."

"Fine."

"I'd also hoped I might drop by today in a short while to have a look at your department."

"I guess today would be as good a time as any," Darren says. "Know where we're at?"

"Actually, I'm in the building now, and I've gotten your location from Angela, so how about if I meet you in a few minutes?"

"I'm in the desk closest to the corner," Darren says.

Five minutes later, Darren and John are shaking hands. John is comfortable that Darren appears to be a pleasant and basically mild-mannered client, and Darren feels comfortable with John, since he appears very businesslike.

After introductions, John asks that Darren give him a brief tour of the department, along with an overview of how things function. In no time at all, Darren is totally at ease with John and going on about how great his entire operation is.

In addition to gaining the information he has asked for, John is quickly deciding that the department would be a workable shoot location. It isn't particularly noisy, it's on the ground floor with easy access, there are plenty of AC outlets on the walls, and a fuse panel is close by. He also makes a quick mental note that the windows face north, and so the sun will never shine directly in—another plus.

As they move through the aisles and among desks, John is keeping an eye out for possible shots and angles and gauging in his mind how much disruption a production crew would cause were he to shoot during a business day using employees as extras.

His other option would be to shoot after-hours, at night, and use hired extras instead of actual employees. These, he knows, run approximately $100 per day, and he figures he'll probably need about five.

Forty minutes later, John is back with Jenny. He notes that it is nearly 5:00 P.M., but he still has a number of topics on his agenda. The two sit down in the conference room again. First Jenny updates John on her progress.

Jenny tells John things are going well. She has pulled all the tapes he requested and cued them up to the actors he wants to view. She has also been told, by Angela, that a new talent agency, Garrison Talent Agency (GTA), is eager to send over actors and actresses to read for parts. GTA is sending a folder of new composites by overnight mail.

She continues, saying that the prop house has given her a price of $975 for the week for the items John has requested. She says she has called another rental place for another bid but as yet does not have prices. She figures that by morning the company will call back and she should have a firm comparison.

"Great," John says. "Now, why don't you call it a day. I'll go over these tapes and leave you notes on which ones to call and some more information on what types I'd like sent over from this new agency. Tomorrow, you can get on the phone and call the client and introduce yourself. Also, let's get the go-ahead from Angela on the cafeteria stuff and line it up tentatively. Then just dig into this breakdown."

"Fine," Jenny says. "Are you coming in tomorrow?"

"After I do some serious scouting. I want to see if I can get all these locations down right away and finalize a set sketch for Angela and the client."

"You still want me to wait on crew?"

"I'll have that worked out tomorrow too, after I get set on the locations. But I can give you these basics now and you could get them lined up."

"OK," Jenny says, getting her pencil ready.

John thinks a moment and says, "I know we'll need a gaffer for the entire shoot. Brian Keen is good, and he works well with Bill. Did you get Bill, by the way?"

"Yes, and he said he's looking forward to seeing you again."

"Good. OK, then we'll also go with Brian as gaffer, and let's get Carmen Obert for sound. She works well in the studio and on location. We'll need a TD for at least 2 days—the tech and the studio shoot. Try

David Miles. I think that's about it for now. We may need a grip and a makeup person, but I want to get a little more into it before I commit on them. The dates, by the way, look like about the 17th through the 24th, not including the weekend. We'll probably build and light on Thursday and Friday, shoot in the studio on Monday, and go remote for the next 3 days."

Jenny has been taking notes. "Sounds fine," she says and puts down her pencil.

After Jenny leaves, John decides to take a look at the talent selections on tapes that she has searched out and cued up. As he is watching them, Angela comes by and sits in on the screenings. Though she trusts John's opinion as a director, she also makes it a general rule to view and approve any talent selections on her programs. This is just to be sure that she gets no surprises.

She and John agree that several of the actors John is checking are possibilities. She tells John to bring in whichever ones he feels are appropriate. While she is there, the two also discuss the cafeteria props rental, the preliminary crew selections, and John's meeting with the client, Darren Whitesall.

When John and Angela finally finish up, it's 6:30, but John feels it has been a productive day. Many initial preproduction items have been addressed, and John knows that if he has a successful day scouting locations tomorrow, he'll be well ahead of the game. With these thoughts in mind, he heads home, has a quick TV dinner, and reads over the script several more times.

16

Location Scouting and Set Design

The next morning, John leaves the house with, among other things, a clipboard for making sketches, a Polaroid camera, a grease pencil, and Jenny's script breakdown. He starts his location scouting with a trip back to Darren Whitesall's office area to confirm that his initial feelings about it as an acceptable location were correct.

He again feels good about the location, and so now he makes a rough overhead sketch (Figure 16.1) and walks among the desks and partitions as he begins to visualize possible angles and coverage. As he does this, he shoots Polaroids of the general area and also refers to various parts of the script.

Soon, a grouping of shots begins to formulate in John's mind. He realizes that he can shoot several of the office shots here, as well as a few of the montage shots. He sees that, with the camera looking in different directions, the backgrounds appear very different, which means he will be able to "cheat" this one location as several different ones.

He looks around for a conference room to accommodate several scenes, but the only one close by is too small. He also scouts Darren's office, and gets a Polaroid and permission to shoot in it if he feels that doing so is appropriate.

Besides the Polaroids and overhead sketches of each of these locations, in every case John verifies three things: the availability of power, the general lighting situation and requirements, and the sound conditions in the area. From these observations, he judges that this office area will be a productive location, and he notes which scenes might go with each Polaroid and sketch. In doing this, he marks the scene numbers on the Polaroids themselves with his grease pencil.

Since the legal department is next to Darren Whitesall's group, John makes this his next stop. He introduces himself to the secretary with a friendly smile. "Hi. I'm John Mitchell," he says. "I'm working with Angela Sparks from the media communications department. We're going to be developing a video program that deals with the governmental affairs group, and two of our shots call for a quick visit to the legal department. I was hoping I might be able to talk with the department head and maybe take a few Polaroid shots."

The secretary's eyes beam when John says "media communica-

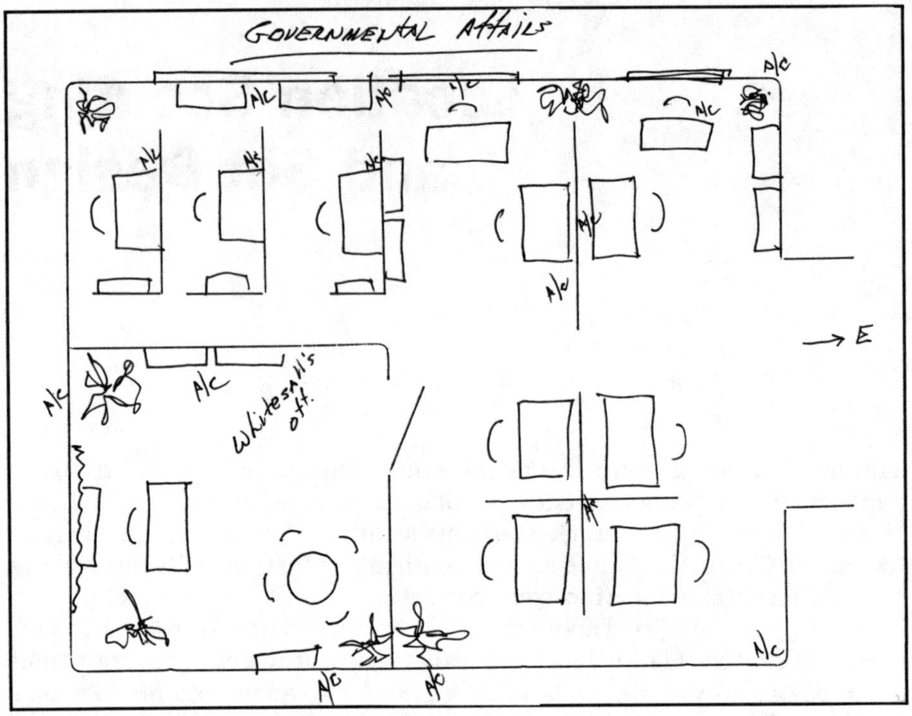

Figure 16.1 Departmental overhead sketch.

tions." She knows what group this is and, like many other corporate employees, hopes she might find her way into a shot or program. She remains businesslike, however, saying, "Let me get Mr. Arlington, for you, Mr. Mitchell."

A few minutes later, the department head approaches and John explains to him in much the same way he did for the secretary. He adds, "I'm directing the program, and I'm a contractor working for Angela Sparks. We'd like very much to get just a shot or two here. I think we'd need a few hours. Here's where it shows up in the script."

The department head reviews the script, sees that it is a minor shot, and gives his approval. Being a lawyer, however, he comments, "I think it will be all right. I'd like to get a copy of that part of the script, though. Oh, and leave me Angela's number also."

"Absolutely," John says, assuming that Arlington wants to run a quick check to be absolutely sure all of this has been approved.

"Will you need my people to be actors?" Arlington asks.

With this question, John sees a nice opening to solidify his attitude of goodwill. "Actually," he says, "we'd hate to inconvenience you, but if there *is* a person or two who would like to help us out, that would be great. I don't mean act in the sense of having scripted lines, though. They'd just have to walk through some natural-looking actions for us. We can bring our own people if you feel yours would be too busy, but it's usually better to get employees involved whenever possible."

Arlington likes this. "Yes," he says. "I think you're right. I'll arrange it."

"Great," John says. "Well, then, if you don't mind, I'll just make a sketch or two and take a few Polaroids. Then I'll have Jennifer Martin, my assistant, call back to get the names and firm things up."

"Fine," Arlington says. "Nice to meet you."

"My pleasure," John says, shaking hands and knowing that this too has been a productive meeting. He then walks through the department, carefully visualizing shots and basic blocking. He also takes his Polaroids, makes his sketches, and does his usual sound and light checks.

Next,it's on to the plant construction department several miles away and a typical housing tract construction site John thinks he remembers seeing close by several days ago. He plans this location for the shots of the superintendent's first meeting with Phylis and later for the phone conversation with both her and Tom.

This location, it turns out, works particularly well because it happens to have a pay phone, which is called for in the scene, at one corner of the lot. Satisfied that everything works, he tries to find the job superintendent to ask for permission to shoot. When he is told the super is gone for the morning, he jots down the construction company's name and phone number, displayed on the job site trailer.

At each location, it has been the same: introductions on a very professional and corporate level, sketches, Polaroids, and power, lighting, and sound checks. John also considers parking at each location and the potential for traffic complications, possible crowd problems, and other such logistics.

With his scouting completed by lunchtime, John stops in at a local fast-food restaurant, gets a booth in the rear, and, over a hamburger and french fries, begins to work out initial camera positions and blocking diagrams (Figure 16.2). He then works out the initial studio set design, shown in Figure 16.3.

The set design shows three sets. The first is his break room set,

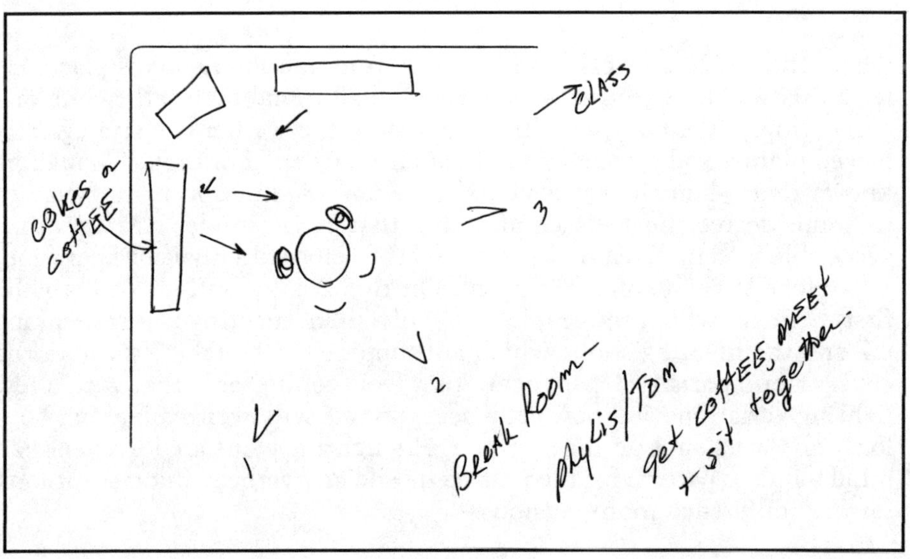

Figure 16.2 Initial camera positions for three-camera studio shoot.

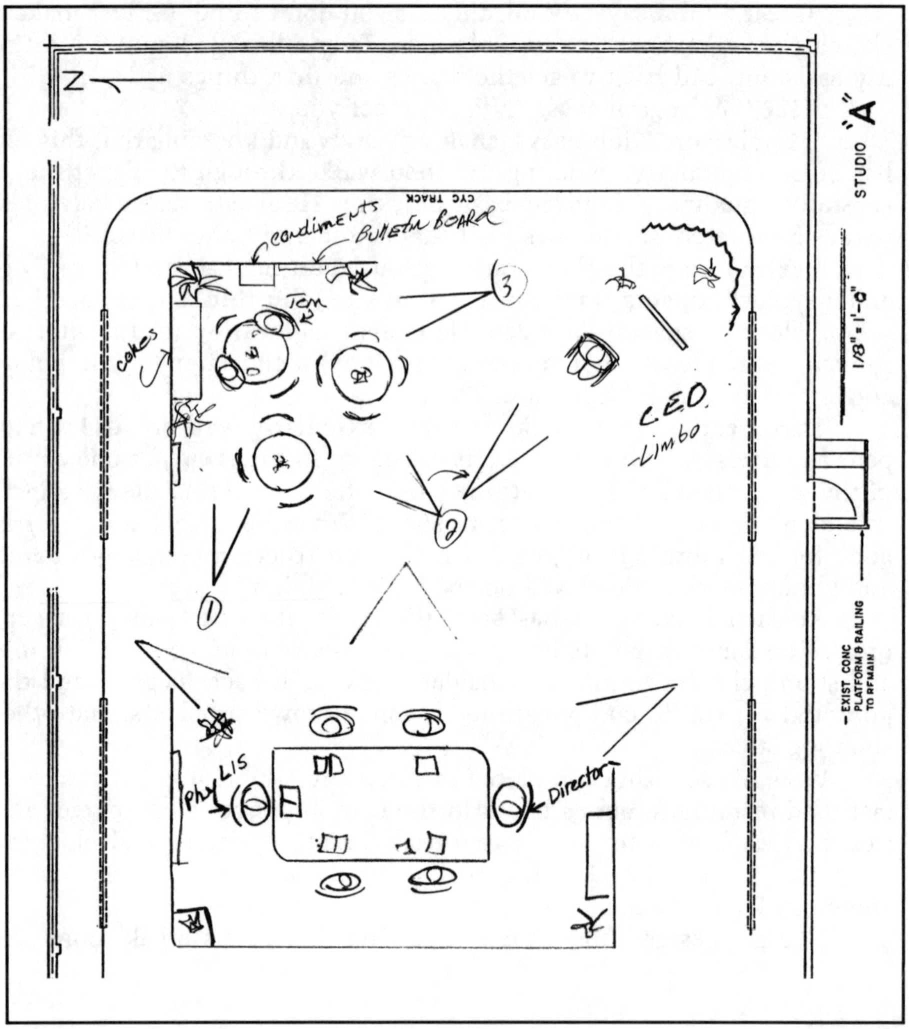

Figure 16.3 Initial set design.

where the entire conversation between Tom and Phylis takes place. He feels this will be effective with several small round tables, the coke machine props, an employee bulletin board on the wall, a few strategically placed plants, and perhaps a small condiments shelf or table. Though he knows that when the set is actually under construction it may change to some degree, he feels comfortable that it is simple, cost-effective, accessible for a three-camera setup, and thus close to a final configuration.

Next is the conference room. On this set, he will shoot both the first scene, in which the employee involvement meeting is just breaking up, and the meeting shots, which are supposed to be taking place in the governmental affairs department. John feels comfortable that with a few lighting variations and new set pieces, no one will ever realize that both locations were shot on the same set. His lighting plans are for a venetian blind window pattern for the first scene and an overhead fluorescent look for the conference room versions.

Finally, he decides he will use a black limbo set with a piece of latticework and still a third lighting look to shoot the CEO shot. This shot, he knows, will be a fairly tight medium close-up, probably done with the CEO sitting on a stool or director's chair. With a plant over one shoulder, a lattice with a decorative slash of light, and typical 3-point lighting, he knows the CEO will look very comfortable and "executivish."

Last but not least, John returns to D&L, Inc. and makes a quick sketch and takes a Polaroid of a pay telephone location in the hallway around the corner from Angela's department. It is here that he will shoot Tom on the phone with his boss, the superintendent.

When he is finished, John has several solid sketches and a good feel for his initial shoot requirements. He begins to sense that familiar feeling that it will soon be time to break each scene down based on time and complexity and then mold this information into a shooting schedule. First, however, he wants to meet again with Jenny to share his progress, check on hers, and look over his audition participants.

Half an hour later, he and Jenny discuss the location and set requirements John's scouting has made necessary. The construction site will require calling the company to gain permission and calling the city as well, for a shooting permit. This, Jenny knows, will also bring up the need for an insurance rider naming the city as beneficiary in case any damage or injuries occur during the shoot. The other locations will each require a call on her part to provide information on the shoot time and logistics, as well as which employees will be local talent.

John also shares his set sketches and discusses these needs with Jenny. He asks for an assortment of trees and plants and for different types of furniture and wall hangings. The smaller items will be borrowed from employees within the department, including Angela. The furniture will come partly from Angela's department and partly from the supply department that handles the delivery of such needs from its local warehouse. Were this local supply resource not available, the furniture, like the coffee shop set pieces, would have to be rented.

Following this, Jenny and John discuss talent. Jenny has a list of people culled from John's tape and composite reviews, as well as a list from the talent agency eager to work with D&L, Inc. She says she has confirmed the auditions for the 10th, and she has 34 actors and actresses. She and John discuss the fact that he will "read" them in pairs whenever possible and, when such pairing is not possible, Jenny will deliver lines as the person opposite the auditioning actor.

Finally, John has a "To Do" list he has put together over the past day. He shares this with Jenny. The list includes such things as a call to the client, just to tell him everything is moving along nicely and provide him with the preliminary shoot dates. John asks that Jenny follow this up by saying he will be contacted very soon about the final dates and times.

John also adds a grip to the crew list for the location days only, since these will require some hustling to stay on schedule, and confirms that the CEO's script has been sent to him for review and possible changes. With this all done, he meets again with Angela and updates her on his location scouting progress. He shares the Polaroids, along with his set sketches, and gains her approval.

The following day, John spends the morning revisiting his locations, finalizing his camera positions, and doing basic blocking. With this information in hand, he knows he will be able to take care of his primary pieces of paperwork—a scene breakdown, a shot list, and a shooting schedule. Before he gets to this, however, he returns to D&L, Inc. to help Jenny set up for the following day's auditions.

They set up a small camera in a conference room and several tables and chairs. John sets up just enough lighting to be able to record the auditions and see the actors' faces clearly.

With that, John heads home and, as on nearly every day since he received the script, rereads it scene by scene, carefully analyzing and interpreting each and every one of its subtle aspects—character types, motivation for each exchange, conflicts, character development, possible camera angles and movement, preliminary blocking, framing, pace and tonal characteristics. He also makes it a point to remain keenly aware of the CEO shot, since, although this one may not involve actors, sets, and dramatic role-play, it is just as important in the producer's and client's estimations.

Finally, before bed, he again reads over the pages that will be used in the auditions and reviews in his mind the types of performance characteristics he will attempt to elicit from his actors and actresses: credibility, range, and ability to take direction.

By 9:00 P.M. he is beat and by 9:15, lost to the world.

Auditions 17

The next morning, the first actress scheduled to audition doesn't show, but her male counterpart does. John considers whether to wait for a female to read across from the actor; then, looking at the long list of people to be seen, he decides against it. He makes sure the actor has roughly 5 minutes with the script pages, or "sides," and then asks Jenny to show him in. He requests that she read opposite him.

John shakes the actor's hand, and their conversation, along with the ensuing audition, takes place generally as follows.

"Hi," John says with a warm smile. "I'm John Mitchell. Nice to meet you."

"I'm Don Marquet," the actors says. "Pleased to meet you too."

"Well, you can have a seat here at the table, Don. Jenny will get a mic on you, and if you don't mind, she'll read opposite you."

"That's fine," the actor says, as Jenny fastens the mic.

"I know 5 minutes isn't much time," John says, "but have you had a chance to get a feel for the script?"

"Yes. I have. Do you mind if I refer to it during the audition?"

"Not at all. But how about if I give you a quick sense of perspective before we roll?"

"That would be good. I get the idea this Tom character is kind of lazy."

"Exactly," John says. "Personally, I also get the sense that he's a bit immature and cocky. I mean, he's positive he can do this governmental affairs job with no effort at all."

"Yes, right."

"And, you know, that makes me wonder about his intelligence level too."

"He wouldn't seem too bright, would he?"

"I doubt it. A lot of the script really kind of revolves around the fact that he just doesn't get what Phylis is trying to tell him—that governmental affairs is a tough job. And when the show ends, he still doesn't realize that these people he thinks have it so cushy have, in fact, put out a good deal of energy actually saving *his* job."

"I get it," the actor says confidently.

"Good," John says. "Well, unless there's something else I can tell you, how about we give it a try. Want a quick rehearsal before we roll tape?"

"I guess that would be good."

"OK. How about you, Jenny—ready to be Phylis?"

Jenny nods.

"Good. Then 'Action,' " John says.

The performance is fair. The actor has gotten an accurate sense of the character, but John feels the delivery is basically flat and inanimate. "Very nice," he says. "I think we should roll. Now, on this one, Don, make it a little bigger and more animated. I'd like to see Tom reflected in his expressions and actions as well as the lines."

"OK," the actor says, still with solid confidence.

John pushes the record button on the small camcorder, waits a second, checks the framing on the camera, and says, "Action."

Again the performance is so-so, and it is a *little* more animated but still basically flat. John decides the actor is a possible, and so he takes the time to give him some further direction. "Again, very nice," he says.

"Thank-you," the actor says.

"But I still feel it needs more animation and projection. Consider the fact that this Tom is the kind of guy who might be pretty laid-back if he's so cocky. And I think he needs to let Phylis know that with facial expressions and gestures. Do you want a minute to think on it?"

"No," the actor says, "I think I see what you're after. You want him to be almost a wise guy."

John thinks about that. "Yes," John says, "let's give that a try."

Again he rolls and says, "Action."

Again the performance is better. The actor has taken direction fairly well, but he still doesn't seem to get what John is after. "OK," John says, realizing he must move on. "That was great. We'll call in a few days and let you know."

The actor gets up, removes the mic, shakes hands, and leaves. Jenny checks out the waiting area—three more people have shown up. The same process continues.

By lunchtime, John has seen enough good actors to convince him he could cast every part but Phylis's. None of the females have been quite up to what he is after.

After lunch, five people are waiting, reading over scripts. The first woman, Ellen Adams, is middle-aged, slightly overweight, but very businesslike in appearance. John likes her look. Jenny pairs her with a handsome young "Tom" named Oscar Billet.

John gives his usual overview and rolls tape. As the pair begin to read, John experiences that wonderful feeling all directors get when they witness a truly credible performance. What happens before him is almost exactly how John had visualized the scene playing out on the set. Both actors are sincere, relaxed, and animated. In addition, John is happy to see Ellen stop and pause occasionally, "thinking" about what Oscar is saying. As she does this, her face tells the story by its look alone. Oscar is equally convincing. He needs only a toothpick in his mouth and his feet up on the table to be a truly cocky, ignorant young fool.

When the performance is finished, John has very little he would ask to be changed, but he wants to check how well each of these actors takes direction. Thus, he says, "Excellent! Very credible. Boy, you guys just nailed it. But I'll tell you what—since that was so easy for you, let's put a little different tone on it, just to see what happens."

Both actors seem eager to hear the challenge.

"Ellen, you did this character just as I would imagine Phylis would—very businesslike and level-headed. But let's turn up the heat on Tom here. Try to *punish* him just a bit. And Tom, instead of the cocky attitude, how about if you're cocky on the surface, but in reality, you're not such a bad guy. In fact, if we listen closely, we can tell you're almost *pleading.*"

"Can we have a few minutes to get into this?" Ellen asks.

Probably a method actor, John guesses. "Sure," he says, "take five." He then leaves the room, gets a cup of coffee, takes a few sips, and returns.

When he says "Action" again, the performance plays out with subtle perfection. Ellen has grasped exactly what he wanted. There is a fiendishly snide sparkle in her eye. Her tone of voice is composed but with just a hint of viciousness. She has become businesslike on the surface but mean and nasty underneath.

Oscar, on the other hand, has become the victim. He is no longer ignorant but slow and empathic instead. He deserves Ellen's mercy but gets only the twist of her dagger.

John sees real character conflict brewing subtly between the two. They are great. "Excellent," he says. "Excellent. You couldn't have nailed it better. I've seen all I need. Thank you both for coming, and let me compliment you on your acting skills. We'll call in a few days."

The actors know John is pleased, and they are too. Both leave feeling confident they will get the parts.

The rest of the afternoon holds more good acting, and when the day is over, John is exhausted but comfortable that he has his cast on tape. He plans to cool off on the performances this evening and view the tape the next day. He will then make his decisions and have Jenny make the callbacks.

After helping Jenny tear down and store the lights and camera equipment, John is heading out the door when Angela approaches. "How'd it go?" she asks.

"Great," John says. "Got some time in the morning?"

"Sure."

"Good. I'd like to show you my choices. Around 10:00?"

Angela appears pleased. Just like last time working with John, things appear well in hand. "See you then," she says. "Have a good night."

18 Talent Selections

By the time John gets into the office the next morning, he has been over the audition performances in his head dozens of times. Although his mind is pretty well made up, he takes an hour over a cup of morning coffee to look over the audition tape. He finalizes his selections and then notes the locations of the selection takes he wants Angela to see. After this, he calls her in to have a look.

Angela likes what she sees. John's judgment is sound, and his choices appear both talented and appropriate in terms of their looks. Her only problem is with the person John has chosen to play the governmental affairs director. Although his performance is excellent, he is a small, thin, balding man. Angela and John discuss the fact that she had envisioned him as more of a heavy, commanding type.

"I had the same feelings initially myself," John says, "but after watching this guy's performance and visualizing him doing the role, I think he'd play just fine."

Angela is not convinced. "Who else did you have for this part—anyone good that fits the bigger image?"

"Actually, yes," John says. "There was another person who was very good. Let's have a look." He finds that audition and plays it. He can tell at once that Angela likes this person much more. The individual is graying at the temples and much taller and broader than John's choice. As John watches him, he feels there is very little difference between the two performances. Both are good, and this taller gentleman also took direction very well. There's no doubt he could do the part.

"Now, him I really like," Angela says. "He really gives some presence to the role. And frankly, I think the client would prefer him also."

"Any value in not going for the obvious choice?" John says. "Maybe giving the role a little something new or different? That was my point."

"I see what you were after, and frankly, I think your small guy would work OK, but in this case I'm pretty sure we'd get a better response from both the client and the audience with more of the typical look."

John sees that Angela is firm in her opinion. "OK," he says, "you got it. The big guy it is."

"Great," she replies. "Do me a favor and have Jenny bump off a copy of just the final selections. Have her leave it on my desk. I'm going to be over in the client's operation later. I'd like to let him have a look. I'm sure he'll be happy with everything, but I think we should get him involved."

"I agree," John says, knowing this is good politics but sometimes problematic, depending on the client.

After lunch, the word comes back: The client approves the entire cast. John immediately tells Jenny to get on the phone and make the bookings. He also asks that she be sure to call the actors that didn't get the parts, or their agents, and thank them for their auditions.

In the same meeting, Jenny gives John some updates. The locations are arranged with one exception. The construction site locations will require a payment of $300, and the city is insisting that, since it is near a busy street corner, a police officer will also have to be arranged for the day. This is an additional $47 per hour, with a 6-hour minimum. Last but not least, the additional insurance rider required by the city is a problem for the company because the city is insisting that it be for $1 million in coverage and the company usually stops at $750,000. This means that a special letter will have to be written and faxed out at once by Angela, to get an approval.

Other than this complication, everything is falling into place nicely. The crew is arranged but waiting for final dates and times; props and furniture are ordered, bought, or borrowed from surrounding offices; and the studio and equipment have been reserved by Angela. All that is required for final confirmation is dates and call times.

John "reads" in Jenny's comments that she is getting antsy for a shooting schedule. He knows this should be next on his list. But first, the issue of the construction site must be addressed. The money is not that big of a problem—$300 a day for a location fee is reasonable. It's the complications it is causing that bother him—especially the fact that Angela will have to write a special letter. He knows this is a pain in the neck for her and that she would probably rather not deal with it on top of all the other problems she is constantly busy with.

Since John always makes it a point to cause the producers he works for the least amount of problems he can, and since he thinks he can cheat the location, he tells Jenny to hold on it until he gets back to her. He also says she will have a shooting schedule by tomorrow.

After this meeting, he takes off for the company construction yard he had visited the previous day. He thought he remembered a corner of the yard that, if shot right, could appear to be a construction site because part of an old concrete slab and some lumber on it were visible. The problem was that it had no pay phone. When John now looks it over, he realizes that a fairly tight camera angle positioned so that the slab and lumber are behind the actor would work perfectly. He also thinks he remembers an old gutted pay telephone in the prop room back at the office. He heads back, confirms that he was right, and tells Jenny to forget the construction site.

19 Schedules and Paperwork

The next morning, John goes to work on his final preparation. He develops a scene breakdown, a shot list, and a shooting schedule.

His scene breakdown is a listing of each scene in chronological script order. The complete handwritten breakdown can be found in Appendix 5. Its basic intent is to lay out the critical elements of each scene, such as locations, talent, and setups, along with any special considerations and a shooting-time estimation.

As John goes through the breakdown, he visualizes each scene and considers the type of coverage he will shoot, the number of times the camera and lighting will have to be struck and reset, the numbers and types of actors, and other considerations or complications that can be anticipated. At this stage of the breakdown, he does not concern himself with driving times to and from locations or break and lunch times.

Once this is completed, John marks the scenes by common locations and then gets out a pair of scissors and cuts the breakdown into individual horizontal strips. He then begins laying the strips together based on his *locations* markings. With this step, the order of his scenes now changes from *chronological* script order to the most efficient *shooting* order. In this, as in most cases with single-camera location productions, the shooting order is totally *out* of order in terms of script sequence.

For instance, though each of the break room scenes appears in different parts of the script, they all take place on the same studio set. Consequently, John now lays these strips one after another, since this is the most efficient way to record them. He gives the same treatment to the conference room scenes.

This begins to shape the entire shoot into large segments of work taking place at similar locations and involving other similarities that will make them cost-effective to shoot together. For example, if he groups his work primarily by locations, which is most often the case, it may require that he use an actor on, say, the first day of the shoot and again on the third day, leaving a nonworking day in between. This is important to note because the actor's union, the Screen Actor's Guild (SAG), requires that its members be paid for these types of days off, since they are restricted from taking other work. At close to $400 per day for each actor, this can have quite an impact on the budget.

The flip side of the coin, however, would be arranging the shooting order by an *actor's* involvement, which might mean shooting more than once at each location. This would be even more expensive, considering that a crew of five might have been at that location totally set up to shoot but had to revisit and reset up the next day to accommodate an actor's schedule.

These considerations and others—such as disruption of the work force, equipment availability, crew availability, and rental fees—have an influence on John's reasoning as he goes about this task.

As the segments eventually begin to gel into cohesive, cost-effective work units, John also adds two 15-minute coffee breaks and a 1-hour lunch break on each day. In addition, he adds driving-time segments on the location days—one for the morning drive to the first location and a second for the drive back into the offices at the end of the day. If there are additional location moves, he adds in this time also.

Eventually, John has before him several sheets of paper with strips taped onto them. These outline a series of logically sequenced shooting days and represent the most economical and effective way to execute the production. A copy can be found in Appendix 6.

With this accomplished, John's first step is done.

His second step is to transfer this information onto a rough, handwritten shooting schedule, a sample of which can be seen in Figure 19.1. The final typed schedule (Appendix 7) is the actual schedule, around which all final arrangements for people, locations, and equipment will be confirmed. A copy of this will go to Angela for her approval and also to the client, along with his invitation to be present on the entire shoot.

*WARDROBE CHANGES

SHOOTING SCHEDULE
FORM PR 2523

V No. 225A TITLE GOVERNMENTAL AFFAIRS

DISTRIBUTION: White - Producer; Yellow - Director; Pink - Ad; Goldenrod - Project File

PAGE: 3

PRODUCER ANGELA DIRECTOR John M. AD JENNY

DATE	DAY	TIME	SCENE	PAGE(S)	DESCRIPTION	LOCATION	TALENT(S)	PROPS
		7:00	-	-	CREW/CAST CALL. LOAD TRUCKS	-	ELLEN	
		8:00	-	-	SET UP FOR 5-2 AND 9-3	LEGAL	ELLEN + EMPL.	
		9:30	5-2	4	PHYLIS VISITS LEGAL - MEETS EMPLOYEE	"	"	PAPERS - NOTEBOOK
		10:00	9-3	7	PHYLIS RETURNS TO LEGAL + (STRIKE LOCATION)	"	ELLEN + SAME EMP.	"
		11:30	-	-	LUNCH			
		12:30	-	-	SET UP FOR 9-2 (CALL TIME PAUL M.)	CONST.		
		1:30	9-2	7	SUPT. ON PHONE W/ELLEN	"	PAUL MUSSER	CLIPBOARD PAPERS
		2:00	9-2	-	SET UP FOR 13	"		
		3:00	13	11	SUPT. TALKS W/TOM	"		"
		3:30	-	-	SET UP FOR 5-3	"		
		4:30	5-3	4	PHYLIS MEETS SUPT.	"	PAUL + ELLEN	AS ABOVE
		5:00	-	-	STRIKE - WRAP FOR DAY			

Figure 19.1 Handwritten page of shooting schedule.

Crew members will also be provided with a shooting schedule, as will the cast. John follows this with a shot list, as seen in Appendix 8, and a package of blocking diagrams like the ones we have previously reviewed.

The shot list itemizes and prioritizes each individual shot. The overhead diagrams provide visual references of where the camera will be positioned to acquire that coverage.

Some directors produce shot lists, and others do not. John has found that doing so is a good idea for several reasons. First, it will mean no extra work for him in production, since he will give it to his AD to monitor and check off as the shoot progresses. Second, it acts as a quick reference guide that has been thought out in advance. This can be a blessing under extreme pressure, when an on-the-spot decision must be made to scrap parts of scenes in order to keep a production on schedule. Third, it gives him a visual tool with which to judge whether or not he is shooting adequate, too much, or not enough coverage. If, as he reviews the list, he finds one scene with 12 shots and another with 2, he might weigh the importance of both scenes as well as other factors that help him determine if this imbalance is correct.

As with the shooting schedule, John also produces the overhead diagrams for quick reference guides. In addition, they act as visual orientation tools for the crew and talent. John and the DP will look over the diagrams as they arrive and begin to set up at each location. The studio diagram will give the DP the same type of visual orientation on the studio sets. The DP may share the diagrams with the grips and gaffer to show positions of lights or other equipment.

Between these three documents, John knows he will provide to all those involved in the shoot a great deal of information that, for the past week, has been assimilating mostly in his head. This information will give everyone a comfortable sense of confidence that the shoot is well in hand and scheduled to come off with skill and precision. John knows it may also bring up questions or concerns that can be addressed prior to the actual shoot.

With this knowledge in mind, John is about to meet with Jenny and instruct her to type out and formalize the shooting schedule and shot list. He will also ask that she reproduce the diagrams and make them available to crew members, Angela, the client, and anyone else involved with the shoot who may want to review them. In addition, he will also tell Jenny that he will review the material briefly for the crew at a preproduction meeting on the stage the first half hour of the morning of the first day.

Before he can do this, however, Jenny shows up at the conference room door. John can see in her eyes that there is a problem.

"I just stationed 12'd the cast," she says, "and Ellen, the lead actress, isn't paid up."

Station 12, John knows, is a process of verifying with the Screen Actors Guild (SAG) that the actors have indeed paid their dues and are in good standing with the union. The station 12 process is a precaution taken to ensure that complications do not arise *after* the shoot that may result in SAG fines or restrictions against the producer and actor.

This news leaves John with two options. The first is to contact the actress and tell her she cannot be cast unless she visits the SAG office at once to take care of any outstanding problems she may have—including unpaid dues. The second is to cancel her booking and recast the part with a second choice. John remembers Ellen's perfectly credible performance and decides to try for option 1. "Call her," he tells Jenny, "and tell her I'd like very much to have her on the shoot. Tell her I feel she'd be a perfect fit for this cast, but we can't bring her on unless she takes care of the station 12 immediately. If there's any question at all, I'll have to recast the part."

"Shall I call up an alternate choice just to check for availability?" Jenny asks.

"No," John responds. "Only if we don't get an assurance that Ellen will take care of the station 12 right away."

John realizes this is a calculated risk. He knows there is very little time to recast the part now, but he is confident that Ellen, like most other actresses he has met, will want very much to do the lead role and will somehow manage to get her SAG issues taken care of. In addition, he knows there were actually two other possible choices for the part, both of whom weren't quite up to Ellen's abilities but who could do a very effective job with the part.

As he is thinking about this, Jenny says, "Oh, by the way, there's one other problem. I almost forgot. The CEO's secretary called, and he has some emergency budget meetings that have just come up. He's only available on Tuesday or Wednesday for his interview. She says she needs a callback sometime today if we want to firm up his schedule."

John looks at his shooting schedule. He had the CEO scheduled for Monday afternoon. He sees, however, that he could possibly move the interview to a very early call before that studio segment.

"If I'm not mistaken," he says, "I think the CEO comes in very early every morning. Call his secretary and ask if we can have him at, say, 6:00 A.M. on Monday. If not, we'll have to just put him in whenever he's available and stop the other stuff to get him recorded. You call. I'll tell Angela."

Jenny had known John would accommodate the CEO's schedule if need be; this is always the case with top company executives. She takes off to make her call, and John meets with Angela. Again Angela is pleased. She feels that John has made two good decisions, with regard to the actress and to the CEO. "Need any help?" she asks.

"Not at this point," John says. "If everything falls into place, we should be OK. If it doesn't, I'll be back."

Half an hour later, the word is in on both problems. The CEO will have to be shot when he can come over—probably on Tuesday at 1:00 P.M.—and Ellen has agreed to take care of her union problem within 2 hours.

John feels comfortable with this. The actress was his main concern. With her committed to the shoot, he knows the performances will be good.

Accommodating the CEO will simply mean stopping the role-play shoot and giving the actors a break. Although John doesn't particularly like breaking the pace of his shoot, he's confident he can return the cast to the same pace and energy level when their shooting resumes.

John knows that at this point his shoot is nearly in place, and he is comfortable that everything looks good. Tomorrow will be a last day to review and finalize details. The day after will begin with a preproduction meeting first thing and, following this, the building and lighting of the set.

He spends that evening and a good part of the next day reviewing his script and making last-minute adjustments.

On the morning of the preproduction meeting, John gathers the crew on the set at 7:00 A.M. After initial greetings, he presents an overview that is typical of his style.

He says, "Before we get started, I want to give everyone an overview of the shoot so we're all on the same track. But before I do that, let me say I'm happy to have you all here. I've worked with everyone here today before, and I know we've got a great crew. The cast is just as good, and we've got a well-written script. So I'm excited about the production. I just hope that, as always, you'll give me your best, and you've got my word that you'll have mine. Now, let's have a look at the agenda."

At this point, John goes over the entire shooting schedule in overview fashion. He talks about the type of coverage he plans to shoot on various scenes and mentions any special considerations that should be noted by the lighting crew or the AD—for instance, hard-to-get dolly shots or scenes in which continuity errors may be a concern.

Half an hour later, when he has completed his introduction, the crew feels comfortable that the shoot has been well planned and thought out and is in good, capable hands—exactly the kind of positive morale John wants to start his shoot.

As a last item, he reviews his set sketches with Bill Olsen and the rest of the lighting crew. He and Bill agree that the break room scenes will have a venetian blind lighting pattern on one wall that will appear to be sunlight streaming in from outside. This will create a relaxed and attractive "morning" look. The conference room set, by contrast, will be a bit darker, for more of a sense of drama, and the CEO set will be an attractive limbo shot.

Immediately following this conversation, the crew gets busy. They bring in and erect "flats" to create walls, they put furniture and props in place, they paint and clean, and Bill, meanwhile, begins to think in terms of lighting. He decides where his key and fill lights will be, what wattage they should be, and what types of instruments should be used—hard lights, soft ones, and so on.

Meanwhile, Jenny is busy with final preparations, such as contracts and payment paperwork for each crew member, contracts and related paperwork for the actors, photographic releases, video shots reports, her master script, and other final calls and arrangements.

As the set goes up over the next 10 hours, a good deal of conversation between Bill Olsen and John takes place with regard to lighting. In addition, studio cameras get rolled onto the floor and powered up. Angles and focal lengths become the topics of more discussion. By late in the day, the sets are mostly in place, painted, and "dressed," and Bill has broken his gaffer away from the construction to help with lighting.

As the day ends, John calls for a brief update from the AD and the DP. Everything appears to be on schedule.

On his way home that evening, John reviews the next day in his mind. The crew will complete the sets in the morning while John reads through the script with his principal cast members. After lunch they'll have a technical (tech) rehearsal on the main set, wherein blocking will be finalized and any final problems ironed out.

That evening, John again reviews his script. By now, he has it and his production strategy virtually memorized.

20 Rehearsal

At 8:00 A.M. the next morning, John shakes hands with his two principal actors—Oscar Billet, who will play Tom, and Ellen Adams, who will play Phylis.

After brief greetings, John shows the two actors to a prearranged conference room. Coffee and a tray of juices and rolls are shared. John has had this small tray arranged to be sure his actors feel comfortable, and important—which they are.

As the group sips and munches, they begin to discuss the script in general and soon start talking over characters. John shares many of the feelings and insights he has gained over the past week of reading and interpreting the script. As his ideas come out, so too do the actors, and a melding of opinions takes place that begins to solidify the characters into their final personalities.

Oscar, for instance, brings up the idea that since the Tom character is young, he should appear impulsive. This is something John has not considered. He visualizes what this might do to the character's performance. "I get the feeling that if Tom were impulsive, he might talk faster and be more animated than Phylis. I see her as very composed and businesslike. What do you think, Ellen?"

"I agree that Phylis should be businesslike," Ellen says. "She represents the mature, hardworking, corporate side of the relationship."

"Right," Oscar says, "and I don't see Tom as having many of those qualities."

"I think you're both right," John says, "but I see a danger here too, and that's to overbake or stereotype the Tom character. I think that if he's too impulsive, he may fit the mold a little too neatly. Personally, I saw him as fairly relaxed—more lazy than impulsive."

"Shall we read a little and see where we go with it?" Ellen suggests.

"Good idea," John says.

With that, the readings begin. As Ellen and Oscar move through the script, John listens. The group stops frequently to talk about an inflection or nuance that should or shouldn't be present, based on their common interpretation of the characters and the story.

For John, this is a very important time. This is the first time he hears how much of his interpretation of these characters reflects in what the actors have brought to the roles. He also begins to consider how he will start to shape the performances as he feels necessary.

For example, Oscar does something that catches John's eye as he reads the lines on page 2:

TOM
So tell me about Governmental Affairs. What exactly do you guys do, locked away up there in the, ah . . . HQ building?

On the final words ". . . ah . . . HQ building," the actor raises his eyebrows in a way that suggests a sly or under-the-table connotation. He delivers the line as if there is some illegal or illicit secret in the "HQ building." John doesn't care for this interpretation of the line. He feels the line should suggest cynicism on Tom's part but not illegality.

Rather than tell Tom how to deliver the line, however, John says, "I'm curious how you feel about this 'HQ building,' Oscar. From your delivery, I get the feeling Tom thinks there're wrongdoings going on in there—corporate payoffs or something."

Oscar thinks about it for a moment and says, "Well, I don't know about wrongdoings, but I think Tom sees the HQ building as a place where things go on that he and most of the rest of the work force are not allowed in on—things like schmoozing."

"I agree with that," John says, thinking about how to communicate his feelings clearly to Oscar. "But," he finally says, "as I visualized Tom, I felt there was more *cynicism* in his opinion of HQ than a kind of sneaky want for the inside scoop, if you know what I mean."

"More negative, you mean?" Oscar asks.

"Right. A subtle suggestion that what goes on in HQ isn't some great secret, but rather just a lot of laziness."

"OK," Oscar says. "I see. Let's try that part again."

He and Ellen reread that page, and this time it works for Tom. The suggestion is cynical instead of sneaky.

Later, on page 9, John has a problem with a subtle transition. The lines are these:

PHYLIS
Yes, Tom. I really do think that if we hadn't done all that, some people really might have lost their jobs. In fact, one of the people I consulted with on the bill was a construction superintendent who probably would have lost half his people.

Tom still doesn't believe it. He chuckles and shakes his head at Phylis's dedication. Phylis now realizes she won't be able to get through to Tom. She takes a last sip of her coffee.

PHYLIS (CONT.)
Listen, I've gotta go, Tom. It was nice talking with you.

Phylis gets to her feet.

John feels that just before Phylis delivers her final two lines, she is making the mental decision that Tom is simply not worth talking to anymore. This is a major turning point for Phylis, but John senses that Ellen's delivery was a bit weak. He decides that her pause was too short. She seemed to move too quickly to the lines "Listen, I've gotta go, Tom . . .," not taking the time to actually make that decision about Tom and let the audience see the wheels turning in her head.

John stops the rehearsal and says, "Ellen, I almost get the feeling that you don't care about Tom at all when you write him off with those last two lines."

Ellen thinks about this and says, "Well, I guess I'm pretty fed up with him. At that point, I just want him out of my sight."

"But I think there's a little more to it than that," John says. "In my mind, Phylis is the type of person who would probably feel pretty *bad* that she couldn't get through to Tom. I mean, I see her as mature and very professional but also caring."

"So rather than just being ticked off," Ellen says, "she's depressed too or kind of sad?"

"Well," John says, "what do you think?"

"I think you're right," Oscar chimes in. "After all, she's spent all this time trying to get through to me. In a way, I think she *does* care about me, or at least about trying to make me see."

"That's it, I think," Ellen says. "She cares about getting *through* to him, not so much about him as a person."

"Right," John agrees. "And I think that when she gives up here, it's really a big thing. I mean, it's really her major turning point in the script."

Ellen nods in agreement.

"So," John continues, "I think you need to take your time with that decision and let us see that it's not easy on you."

Ellen understands. "OK," she says. "Let's try it again."

This time, the read is much better. John compliments Ellen on her ability to grasp this subtlety, and Ellen is happy that she is pleasing John.

For the rest of the morning, this type of exchange goes on as the group covers the entire script. In the end, it becomes a process of focusing in on the characters in the script and reaching a common vision of how each one thinks, feels, and acts.

By lunchtime, John feels that a common vision is nearly in place. He is sure that while blocking the scenes that afternoon, the finishing touches will be put in place.

After a relaxed lunch together, the cast and crew report to the stage for the tech rehearsal. The set is built and lit and three cameras are in position on the floor, in accordance with John's overhead diagram.

John asks all the crew members to now take their positions and put on their headsets. As the crew members are doing this, John walks through what he has decided will be his tentative blocking. Since most of the conversation between Tom and Phylis takes place while the two are seated at a table, the main point of discussion regarding camera movement is their entrance and exit.

"I'm thinking," John says, "that we'll start with you just stepping away from the coffee machine, Oscar, and you coming in from the left over here, Ellen. We'll keep that in a wide shot on camera 2. Then, as you both take a seat, we'll push in to a closer two shot and start switching to singles as you talk. Let's walk through that once."

As the actors walk through this blocking, John first watches from the studio floor. He then moves into the control booth and begins to get a sense of when camera 2 should zoom in to the two shot he had mentioned.

After a run-through, John comes on the PA system and asks that the cast hold for a moment. He then gets on the private line (PL) and talks to his DP, Bill Olsen.

"OK, Bill," he says, "this time when Oscar starts away from the machine, let's push in so that by the time he sits down, we're just arriving at the two shot.

"OK," Bill says.

John gets back on the PA system and says, "OK, we're going to walk through that again and look at the camera move." The actors step back out of frame, and John says, "Action."

As the scene plays out, John is not completely comfortable with it. He begins to ask himself why. The actors seem to be doing a nice job, and the camera move is smooth. After some consideration, John decides that while the action seems natural, the elements in the frame appear off kilter somehow. Finally, he decides it is the position of Ellen in the foreground, along with the large coke machine just behind her, as opposed to Oscar in the background, at the coffee machine. This creates an aesthetic imbalance in the frame that makes it appear cumbersome-looking.

John tries having Ellen's entrance come from the opposite side, but this doesn't seem to help. Finally, an idea occurs to him. He leaves the control booth and walks out onto the floor. "This isn't working," he says, "but I've got an idea: What if we eliminate the entrance? We'll just start with a close-up of the tabletop. As your cup of coffee is placed down into the frame, Ellen, Bill will dolly back. You'll just be sitting down as the frame widens out, and Oscar will be stepping up to have a seat."

Everyone agrees this is worth a try. As the cast and crew again walk through the shot, this time John feels it works fine. The camera's move back is motivated to reveal the two just having a seat, and with Oscar just taking a seat the frame is balanced and pleasing. In addition, John visualizes how this shot will cut together with the shot just prior to it, and he feels it will be a smoother transition. This is because the prior shot will be a fairly wide two shot and this, being a close-up, will begin on a substantial change of focal length.

With this worked out, John now begins to check out the camera positions and focal lengths of the other two cameras. He switches to each shot and looks over the framing and composition. He instructs the camera operators to adjust their positions and focal length so that each shot will provide the best framing and composition possible. During this process, several lights must be adjusted to eliminate hot spots and unnatural-looking shadows. Final camera positions are "spiked" or marked on the studio floor with gaffer's tape.

A microphone problem also arises. As Oscar is being seated at the table, the wireless RF mic taped under his collar rustles, creating a distracting awareness of the microphone's presence. After a discussion with his audio recordist, John decides to do away with the wireless mics and go with a shotgun mic, positioned just above the two actors out of frame, on a "C" stand.

This change of mics is possible because John's opening shot no longer requires Oscar to cross to the table. In John's first blocking, Oscar might have begun his line while still moving from the coffee machine to the table. Since the shot will begin in a close-up as the coffee cup is placed down, the line can wait until Oscar is sitting down just under the shotgun mic.

The remainder of the day is spent on these types of adjustments and on segments of actual rehearsal, including John's switching cameras live, as the performances play out. By 6:00 P.M., John, the cast, and the crew are comfortable that all of the bugs are worked out and that the following day will be an organized and productive one.

This shared sense of confidence is a frame of mind John has purposely tried to create. It is important to him because he knows that a group feeling confident and organized will begin the shoot on a positive note. Part of his job over the coming days will be to keep that positive morale going.

Studio Production

21

At 7:00 A.M. the following morning, the studio is coming to life. The TD has loaded a VTR with videotape, turned on the audio system, and fired up all three cameras. He has begun timing signals and adjusting an assortment of other related equipment for the shoot.

Angela has had coffee and rolls brought in for the client and his assistant. These two have just shown up and are milling about on the set. Jenny is seeing to any needs that arise, including those of the client, crew, cast, and director. Crew members are readying their cameras and work positions. They adjust the friction of the camera heads, double-check tape spikes on the floor for dolly moves worked out at the rehearsal, line up shots, and discuss lighting and blocking among themselves.

John has visited the dressing room and approved the actors' wardrobes. The cast for the morning includes Ellen, Oscar, and five extras. Each actor has brought several changes of clothing, based on what John has requested.

The day before, at the rehearsal, John had made his basic choices for Ellen and Oscar. But now he inspects each actor in makeup and wardrobe to be sure everything is perfect.

By 8:00 A.M., everything is set. The crew is on the studio floor, as are the actors. The client is seated in the control booth with Angela and his assistant, and John is on the floor beginning his final walk-throughs for scene 3.

As the actors go through their initial paces, John likes what he sees. The rehearsal time has paid off. Oscar and Ellen are comfortable with the scene, and they have obviously grasped the interpretation John has of the two characters. In addition, the extras fill in the surrounding area with a natural sense of movement and human presence that gives the scene a truly credible look. Just as he has blocked the movement of Oscar and Ellen, John also choreographs how the extras will enter, exit, and sit at nearby tables.

Oscar suggests that when he has a seat across from Ellen, he place his feet up on a nearby seat to reinforce the idea that he is lazy. John had originally envisioned Oscar flipping the chair around with the back forward as he was being seated, but he likes Oscar's ideas and agrees to try it.

After two walk-through rehearsals, John turns to Bill, who is operating camera 2. The DP has been practicing the dolly back with each rehearsal. "What do you think Bill?" John says. "Working OK for you?"

"Fine," Bill says.

"OK," John returns, "then let's rehearse with cameras."

With this, he heads for the control booth. Once inside, he turns to Angela and the client as he is having a seat at the control console. "Looking OK?" he asks. Both Angela and the client are pleased. They give the go-ahead.

John has a seat and puts on his headset. With this, he is connected to the entire crew on the PA. He asks, "Everyone ready?" Jenny, on his right, acknowledges, and his TD, on the left, says, "Let's do it."

"Camera 1?" John says.

"Ready," comes the response.

"Camera 2?"

"Ready when you are."

"Camera 3?"

"Ready."

"Floor manager?"

"Ready on the floor."

John now refers to his script, which is marked with camera "take" numbers, as shown in Figure 21.1. The numbers indicate which cameras he will switch to as the performance plays out. "Let's rehearse one," he says. "Bill, if you could go to the opening close-up, please."

Bill dollies his camera in and tilts it down so that it is looking at the tabletop. "How's that?" he says.

"A little looser," John says. Bill zooms out slightly, creating a wider frame.

"Oscar, will you place your coffee cup into the frame on the table once for me?" John asks over the PA system. He says this because he wants to see how the shot will look as action begins. Oscar does as John asks.

"OK, that looks fine. Thank-you," John says. "Camera 2, when the cup hits the table, that's your cue to dolly."

"Ready," Bill says.

John turns to his TD. "OK," he says, "we're starting on camera 2 and going to 3."

Working with the switcher, the TD places camera 2 on "program," which means it is the signal being sent to the VTR, and camera 3 on "preset," which makes it *ready* to go to "program." "Ready," he says.

"Dale," John says into his headset, "count them down." Dale is his floor manager. Hearing this, he says, "OK, count down for rehearsal. . . . And in 5 . . . 4 . . . 3 . . . 2 . . ."

A moment later, the actors begin. As John watches the scene play out, he also calls his camera switches to his TD. The TD "takes" those cameras on command.

When the scene finishes, John is basically pleased, as are the client and Angela. The performances are good. The dolly back from the table is a natural way to open the shot. The set looks great, as does the lighting.

"Three things," John says. "Camera 2, start your dolly the instant Oscar's coffee cup hits the table. You waited a second that time, and I think the scene will lag a bit if we don't come out to the wide shot right away. Also, camera 3, on Ellen's last line, I'm going to cue you to do a

DLTV - 225A 2

TOM
Listen, I'm in the process of looking for a transfer. I've about had it with my job. Could I maybe pick your brain a little?

Phylis gets her cup and moves to a nearby table.

PHYLIS
Sure.

TOM
Great.

Tom gets his cup and moves the table. He takes a seat.

TOM
... I'll tell you, Phylis, this field work rat race is really crazy. Foul-ups, deadlines, customer problems. I'd like to get into something a little more...
(Thinks)
... I guess "controlled" and "relaxed" is the word.

Phylis shoots a quick glance at Tom after the word "relaxed."

PHYLIS
I see.

TOM
So tell me about Governmental Affairs. What exactly do you do locked away up there in the, ah ...
(Smiles)
"H. Q." building?

Again, Phylis glances at Tom. Her expression tells us she's not happy with the attitude reflected in Tom's comment. She does answer, however.

PHYLIS
Well, basically, Tom, there are a lot of things happening in the state government every day that can have a big effect on the company.

Figure 21.1 Script page marked for live, three-camera shoot.

slow zoom in to a close-up. This is a transition to a flashback. I think if we're pushing in, a dissolve back in time will work well later in post. But it has to be fairly slow so that the dissolve can happen *during* the zoom."

"You got it," says the operator on camera 3.

John says to Betty, an extra, "Betty, try to cross to your table a little sooner. If you linger there by the machine, it becomes distracting."

"Fine," Betty responds.

"OK," John says, "Let's get one on tape."

The TD puts the VTR into record. Several seconds later, he says, "Speed," indicating to John that the machine is recording and that it is safe to proceed.

John says, "Count 'em."

The floor manager does his count again, the scene begins to play out, and soon John is saying, "Good. . . . Ready 3. . . . Hold the dolly. . . . Take 3 . . . , ready 1 . . . , take 1, ready 3, take 3 . . . 1 , give me a medium close-up . . . good. . . . Ready 1 . . . , take 1 . . . 3, go to the close-up. Ready 3. . . . Hold there. . . . Take 3 . . . ," and so on, until the scene ends.

"Cut," John says, finally. "Let's have a look at it."

As he says this, the floor manager tells the crew and actors to relax for a few moments because the shot is being reviewed. Meanwhile, the TD rewinds the VTR and plays back the scene.

It looks generally OK but contains several problems. The client comments that Oscar's performance is a bit off. "He's coming off *too* lazy and wise," he remarks. John agrees. Also, the dolly back appeared rough and wobbly, and the zoom in on the last line was too fast. It ended before the point where John visualizes the dissolve to take place. On top of all this, Jenny says, "Ellen missed the phrase ' . . . can have a big effect on the company.' She said only the first part of the sentence."

The client was watching his script, as was Jenny, and he confirms that he heard this too. "We've got to have it in there," he says. "That's an important point."

John tells Jenny to prepare for take 2 while he visits the floor. He then leaves the control booth, walks out onto the studio floor, and says, "OK, that was a good start, but we've got a few rough spots. Bill, your dolly had a bump in it or something."

"I know," Bill says. "It was a catch in the friction. I'll get it next time."

"Good." John turns to Allison, the camera 3 operator, "Your zoom was too fast, Allison. I'm after a nice, slow push so I can dissolve in the middle of it as Ellen is remembering."

"A lot slower?" the operator asks.

"About half the speed you were going," John says.

"OK."

"And Betty, your walk was fine."

John now moves to the table where the actors are seated. To both, he says, "Very nice. It looked natural and pretty smooth, but we've got a few rough edges. Oscar, on this next one, try to pull back a little on your laziness. As viewers, I think we deserve to maybe be a bit upset with you, but not dislike you."

"I'm being a *real* jerk, huh?" Oscar says.

"Just a little too much. Besides, I don't want to give your personality away in the first scene. I'd like the audience to come to realize what you're like. So let's ease into it a bit more."

"No problem," Oscar says.

"Anything I can do differently?" Ellen asks.

"No," John responds. "You're right on, Ellen." As John says this, he is careful to recognize Ellen's work but not to sound so enthusiastic that Oscar will feel dejected. John will be cognizant of this balance between the two actors throughout the shoot.

As he is about to step away from the actors, Jenny corrects John over the PA system. "Ah, Ellen missed that line, John. Did you want to mention that?"

"Which one?" Ellen asks.

Jenny comes back on the PA and recites that part of the line Ellen missed. Ellen checks her script, acknowledges she forgot it, and promises to get it on the next take.

John then returns to the control booth and take 2 is recorded. This time, Oscar gets the performance right and the initial dolly is smooth, but the ending zoom is now too slow. Also, Ellen misses her line again, and John misses a switch, calling out the wrong camera at a critical spot.

John again visits the floor, discusses problems in the scene, and returns to the booth. On take 3, both Oscar and Ellen miss certain lines and stumble uncomfortably at several spots.

The client says he doesn't care for the overall tone of take 4. This worries John slightly, because some clients begin to play directors when they get into things like tone and visual aesthetics. He agrees with the client, however, and goes for take 5.

On 5 there is a false start—Oscar begins, blows his line immediately, and asks to start over. Take 6 is almost perfect, but an extra looks at the camera for an instant, and Ellen seems to pause awkwardly at one point. John is tempted to buy it but, instead, says over the PA system, "That one was a keeper. It was close. There were just a few rough edges. Nothing I can nail down exactly. We just need another time or two to get it 100%."

Although the cast and crew want John to accept what they have done, they also respect his tenacity in waiting for that perfect take. On take 7, there is another false start, and take 8 is clumsy throughout. On take 9, however, the scene finally plays out beautifully.

"Let's see playback," John says.

The TD plays back the scene, and John sees only two very minor problems: One of his switches was a fraction of a second late, and the end zoom had a slight wobble in it. John checks his watch. He knows he needs to move along, and he knows the cost of 100% perfection is expensive in terms of time and resources. The late switch, he figures, can be tightened in editing, and the wobble in the end zoom won't be a problem if he can dissolve out of the scene soon enough. "Let's see it again," he says.

The scene is replayed, and after rechecking the end zoom, John says, "We'll buy that and move onto scene 6."

Both Angela and the client agree with John's assessment, and neither says a word. Jenny circles take 9 on her master script, and John leaves the booth to discuss the next scene on the floor.

This pattern repeats itself throughout the early morning. At 10:15 A.M., however, a call comes in to the booth from the department secretary. The CEO has just left his meeting and is on the way over for his taping.

John immediately goes to the floor and gives the actors a break. He then tells Bill to drop the lights on the coffee shop set and fire up the lights on the interview set. The floor manager now takes over as a teleprompter operator, and the CEO's brief script is projected on a translucent glass pane in front of the camera.

Five minutes later, Martin Crain, the CEO, steps into the studio with Angela beside him. "I'd like to introduce John Mitchell," she says. "He's our director."

John shakes Crain's hand and says, "Well, we're all set. If you'd like to have a seat right here, we'll just get a microphone on and we'll have you out of here in a few minutes."

"Good," Crain says dryly. "I've got a plane waiting."

The sound recordist places a small lavaliere mic on him, Bill does a quick adjustment on the backlight, and John says, "Your lines will come up right here on the front of the camera."

"I know," Crain responds. "I've done this a few times."

"Good," John says. "Then I'll just head for the booth, and we can rehearse one time and then record it."

"Forget the rehearsal," Crain says. "Let's just do it. I've got to get a move on."

John doesn't like the sound of this, but he heads for the booth, saying, "Yes, sir."

Once in the booth, he puts on his headset immediately and says to Bill, "OK, give me a medium close-up. Leave enough room for an ID title at the bottom." Bill frames up the shot. John asks the CEO if he's ready and gets an affirmative response. "Roll tape," he says to the TD.

When John gets the "speed" confirmation, he instructs the floor manager to count down the CEO. He does so, and Crain reads his lines—very poorly.

As soon as he finishes, John comes over the PA with, "That was a good one for starters, sir, but I'd like to do another one if you don't mind."

Crain grumbles and says, "OK, I guess."

The second take isn't much better, but John can see that Crain is getting antsy. "I know how busy you must be, sir," he says, as diplomatically as he can, "but I think we need just a few more times through to smooth it out a little."

The CEO chuckles, "It's not going to get much smoother," he says.

John knows he is right. CEOs, just like other company executives, are not actors and rarely do a professional job when addressing the camera. John also realizes, however, that at this moment he is making a decision that will help shape the CEO's image with many employees in the company. If he lets a shoddy performance get by just to get Crain out of the studio quickly, later it will be frowned on. On the other hand, he knows that he has only a short time before Crain gets downright angry—which is also something he must not let happen.

The next take is much better. When he's finished, Crain says, "How's that for smooth?" John can read a definite edge of impatience in his voice. Angela is also nervous. John glances at her, and she nods, as if to say that was probably the best they will get.

John wants to buy a little time. "Very smooth," he says over the PA. "But I'd like to see that played back. If you could give us just 30 seconds."

The CEO fidgets but agrees.

The playback is good, but John thinks one more time will produce a better one. He turns to Angela and says, "Angela, that one will play, but I really think he can do better with another take or two."

Angela is nervous but says, "Go for it."

"That was a good one," John says over the PA, "but, sir, we want to be positive that everything is perfect when employees see this. I'm sure that if you can give us the time to do one or two more takes, we'll have a perfect one."

Crain grumbles again but says, "OK. But let's do it—two more and that's it. I'm late as it is."

The next take is also good, and John says, "OK, that was a good one; I only want one more for insurance."

This take is the best yet. It works. John is happy that he pushed for a bit more from the executive. Moreover, now that the CEO is off the hot seat, so is Angela.

"Perfect," John says and heads for the studio floor. He shakes Crain's hand as his mic is coming off and says, "I do appreciate your patience. I know how busy you are. But I think the extra time got us a very nice piece to close this program. Thank you, sir."

John can see that the CEO is impatient but not angry. He too must have realized that the last one was his best. "Good," he says. "I'm glad you're happy with it." With that, he is ushered out the door.

John looks at his watch: 10:40. Two more scenes to get out before lunch.

The remainder of the morning goes smoothly, and the break room scenes are wrapped up by 12:15.

Just before lunch, the actors in the conference room scenes—scheduled to be shot after lunch—arrive, and John meets with them briefly. He suggests wardrobe and tells them they will be having lunch before starting their scenes.

An hour later, the crew and cast are back to work. Since these scenes were not previously rehearsed, working out blocking and performances takes a bit more time. There are walk-throughs, discussions about performance characteristics and energy levels, and cues for zooms and dollies.

John is prepared for this, however, since he has worked out detailed blocking and camera diagrams prior to the shoot. So now, rather than *inventing* what is to happen on the studio floor, his work becomes a matter of explaining what he as a director has previsualized, walking through it, and then rehearsing and deciding if it works. In most cases, it does, and only minor adjustments to camera moves and performances are needed.

After this, just as on any other shoot, it is simply a matter of repetition until everything happens right in one take.

The day ends with everything on tape John had hoped for. The crew takes the last half hour of the day to prepare the equipment for the following day's location shoot.

As they do this, Angela, the client, and John look over playback of several of the buy takes for that day. All agree that the material recorded was very convincing and was extremely well lit and shot.

At 7:00 P.M., John heads home for a relaxing bath and a look over tomorrow's scenes and shooting schedule.

22 Location Production

The next morning, the crew shows up at 7:00 A.M. The equipment is ready to go. The cast call time is also 7:00, and they too are on time and ready to go.

Since the day's location shoot will be at the governmental affairs offices in the same building, the equipment can be loaded on dollies and rolled down a hall, up one flight in an elevator, and across a large employee work area.

John first meets with the cast to discuss wardrobe, the day's shooting, and any questions they may have. He then leaves them in Jenny's hands and walks the crew along the route they will take to governmental affairs. When they reach the area, the crew members are sent back for the equipment and John and Bill, the DP, talk over the day's shooting.

This day will consist of all the shots requiring company office environments. It includes scene 4, the last part of scene 5, scene 7, and part of scene 9.

Since several of these scenes must appear to be different locations, John has set up a blocking diagram that calls for the camera to stay in one general area but to face in different directions for various shots. This is called "cheating" locations. After editing, it will give the viewer the impression of totally different places.

For instance, the location for scenes 4 and 7 will be Phylis's desk area in the governmental affairs office. The location will be given a different perspective in scenes 5 and 9, where Phylis is seen on the phone, and it will also appear, in these same scenes, to be the plant construction department.

After a discussion of the blocking diagram, the shooting order, and the type of coverage John wants to get, Bill is left with a good sense of where to place lights and equipment for the least amount of resetup between scenes. When the crew members arrive with the equipment, he instructs them where to place lights, electric cables, the video monitors, and other equipment.

As this is taking place, Jenny arrives with Ellen and Chad, the actor who will play the governmental affairs director. The call times for the remaining cast members is after lunch.

John now has a brief discussion with the sound recordist, wherein he explains where the camera will be placed, what the action will consist of, and what type of coverage he is after. This information gives the recordist what she needs to choose the right microphones and prepare them for the shots.

Once this is done, John begins talking with Ellen and Chad about general blocking. It is during this conversation that the client, Darren Whitesall, approaches with his script in hand. John senses something is wrong.

"John," he says. "I hate to tell you this, but I got a call this morning from my vice-president. He didn't get to review the script until late last night, and he wants a few changes."

John gets a sinking feeling in his stomach that tells him the changes will be on areas *already recorded.* Looking over Darren's notes, John realizes he is right. The VP wasn't 100% happy with some of what Phylis said in scene 12, and he wants to revise Tom's dialogue in scene 10.

After pulling out his shooting schedule, John instructs the crew to continue their work as he calls for a powwow with Jenny.

They look over what the changes entail. The revision to Phylis's dialogue, they feel, can be squeezed in later that morning with a voice-over pickup, since they will be doing some voice-over work anyway. The fact that Phylis will not be on camera for that shot will require that later, in editing, a cutaway reaction shot of Tom be used to cover the picture portion.

Although this doesn't make John particularly happy, it is at least achievable, with little effect on the shoot. The only real problem, he foresees, will be to match the audio that was recorded yesterday in the studio. He talks to the sound recordist and is assured that if they record the voice-over in a quiet conference room and use the same shotgun mic, the dialogue should be close enough to match.

The revisions to Tom's material are more of a problem, since Oscar will have to be called back in and rescheduled.

"Shall I have someone call him now," Jenny asks, "to see if he can come in this afternoon?"

John thinks this over. The schedule is tight as is. Adding another pickup will take even more time. He looks over the changes and decides he has no choice. "Yes," he says, "but call Angela, and have someone over there do it. I need you here. Tell them to let us know as soon as they reach him, and tell them late afternoon is best. We'll be doing the voice-over work at 3:30."

Jenny immediately leaves to make the call, and John goes back to work on the scenes at hand. He knows that he must remain fully composed and focused in order to keep up his cast and crew's confidence. He is also consoled by the fact that Jenny is excellent at handling such snafus and that Angela is the type who will try very hard to see that John gets the support he needs.

By 8:30, scene 4 is lit and ready to go. John walks through it with Ellen and Chad and decides it is ready to rehearse.

Since he is now shooting single-camera style, his strategy for recording the scene is to first get a solid master scene, including all the action. This will be a hand-held shot that starts almost dead-on to Phylis sitting at her desk and then moves quickly around to the side to include

Chad as he enters. It will then zoom in on Chad to a medium close-up during his last line, leaving Phylis off camera for her last line.

John feels that the slight motion created with a hand-held shot will add a sense of urgency to the shot, as will the push in to Chad during his line. He also feels that having the camera at eye level with Phylis will work well, because when the zoom in on Chad takes place, this will put the camera *below* his eye level, since he is standing—another subtlety that should increase the sense of tension.

After recording a good master, John plans to bring the camera around for a close-up of Phylis doing her lines, including her reactions to Chad's lines. This, he feels, should cover the scene adequately.

His preference will be to use the close-up of Phylis only for the last line. But getting all three of her lines, plus her reactions in a close-up, gives him additional options in editing should he need them.

The first rehearsal looks good, but John is not happy with Chad's performance. He feels the actor is coming across too casual about dropping the papers on Phylis's desk. Although his action is not a major problem, John decides he needs to deliver a businesslike performance versus an informal one.

When the rehearsal is completed, he approaches the two actors. "OK. Good. I think we can smooth it out a little with one more run-through. And Chad, do me a favor, keep in mind that you're a corporate 'lifer' here, so to speak. You're strictly business."

"Isn't that what I did?" the actor asks.

"Basically, yes, but I also felt you were a bit too casual. I mean, I realize that giving Phylis some work to do isn't a big deal, but I think that at your level, you would probably project just a little more of a formal, executive-type image."

"I disagree," Chad says. "I think that, a lot of the time, corporate executives are more informal than we think. The problem is, I think, that lots of people don't really understand that."

"You may be right," John says. "But I don't want this audience to have to take the time to figure that out here or even to think about it. I'd prefer we keep you projecting the traditional image."

John can see that Chad isn't convinced. "Another reason," John continues, "is the fact that we want this audience to gain a real sense of respect for this department. We don't want them to even *begin* to think you're not running a real tight ship here."

Again Chad disagrees. "But don't you think they'll respect me more if I'm casual and communicating with my people on an informal level?" he asks.

"Maybe," John says, "but I'm not ready to take that chance here. I want to be *positive* of how they perceive you. I think it's very important to our overall message."

"I think you should take the chance on it," Chad says.

John smiles. He is getting irritated. "Maybe on the next one," he says with a smile.

"I'll tell you what," Chad continues. "Let me try this, and we'll see what we get."

John now realizes it is time to have a brief chat with Chad. The actor not only is being stubborn but also is beginning to place himself in the director's chair. John knows there is room on any shoot for only one director, and while he is always open to suggestions, he is still the boss. When he feels strongly about something, as he does in the case of this performance, he requires that his cast and crew give him what he needs—especially since the schedule is tight and they will have to squeeze in unscheduled voice-over pickups later in the day.

With this in mind, John walks around the desk directly in front of Chad, looks him squarely in the eyes, and says, "I understand that you may have different feelings about this scene, Chad, but let me be clear on something. We're on a very tight schedule today, and I feel strongly about this performance. If you want to discuss executive personalities, we can do that later, during lunch, maybe. But for now, as a member of this team, I need you to deliver me the performance I'm after."

Chad sees that he has ruffled John's feathers. "Fine," he says. "No problem, honestly. I just felt I should be frank about this."

"And I can appreciate that," John says, "but we need to get rolling for now."

"Hey, you got it, boss," Chad says with a smile.

John moves away from the desk, saying, "Good. Let's rehearse again, please."

The rehearsal looks good. Chad is giving John what he is after. John appreciates his professionalism, though he is still a bit irritated by the incident.

Tape rolls, and the recordings are made. Takes 1, 2, and 3 have camera problems—momentary losses of focus and framing problems. John knows that hand-held work is not easy, and he has expected this. The next few takes are better, but in one, an employee walking through the background stops and looks at the camera, and in two more, the actors blow their lines.

Take 8 is the buy.

The crew moves on to the close-up of Ellen and then the other scenes on the schedule. Word comes in that Oscar can indeed come in in the afternoon for the voice-over work, and he does so.

Everything goes basically well, though the pickups and other minor problems have resulted in half an hour of overtime. The day ends at 6:30 P.M., as the last bit of equipment is rolled back into the equipment room in Angela's department.

John, the DP, and Angela spend another hour looking over dailies. Everything looks good, and John is home, exhausted, by 8:15. By 9:30, he is asleep.

The remainder of the shoot goes according to the same basic pattern. The group arrives at a location and, as John and Bill discuss the shots, the crew begins to ready equipment. Jenny works with the actors and client. Rehearsal takes place, and, eventually, recording begins. In almost every case, the problem for John is the same: getting everything he wishes to happen—solid performances, quality visual aesthetics, quality sound aesthetics, and so on—*all in one take.* Typically, if one element is right on, another will be a little off. Balancing his desire for perfection against the time and money available is a trade-off John is making constantly. Having the ability to make the proper decisions under those constant pressures is what Angela is paying him for.

On the last day, the crew members are ahead of schedule, and John feels they will be in unloading equipment for the last time by noon. His hope for an early wrap, however, is not to materialize. At the last location, Jenny comes with a message from Angela. Please call her ASAP. She would like some interviews shot.

John calls Angela immediately. She explains that the client has just called her and asked for a favor. He has said that although he realizes it is much too late to be bringing it up, in his department are a number of employees with whom he has worked in the past and who would be willing to do testimonials on how hardworking his people are. He has asked if it would be possible to quickly line up and shoot some of these people.

Although John doesn't like surprises like this, he is glad it comes with a half day left to spare. He is also aware that employee testimonials often add a level of real credibility to corporate programs. Thus, he says, "Tell you what. If you can get them lined up for this afternoon, there at your place, we can do them."

Angela says, "Leave it to me."

Shortly after lunch, John and Bill develop an impromptu interview set on the stage. They decide on an attractive background lattice, a plant, a slash of light on the cyclorama ("cyc"), and mildly contrasty lighting.

As the crew gets busy building and lighting this, John visits the three employees to be interviewed. They have been asked to wait in the actor's dressing room, and when John shakes hands with each of them, he makes a quick mental assessment of each. The first two seem fairly comfortable and easygoing, but the third looks extremely nervous.

John knows that the keys to effective employee interviews are getting the subjects to relax and asking the right questions. Toward that end, he starts in his usual, relaxed manner with an overview.

He says, "Well, is this as much of a surprise for you three as it is for me?"

All three acknowledge that they too have been "drafted" on the spot.

"I thought so," John says. "Well, I'll tell you what can make it easiest on all of us—just doing this like a series of simple news-style interviews. Have you seen how people on the street are interviewed on the news?"

Again the three nod.

"Well," John continues, "that's what we'll do. We'll basically just have a brief conversation about the governmental affairs department. We'll have the camera there, but you can forget about it and just talk to me. We'll edit out any flubs and pauses later, so you don't have to worry about getting it perfect. In fact, keep in mind that in most cases, we'll end up using only a few *seconds* of what you say, even though we may talk for 10 minutes or so."

"So we don't ever have to look at the camera?" the nervous one asks.

"No," John reassures him. "All you have to do is talk to me."

"What will you ask us?" another one asks.

Thinking off the top of his head, John says, "Probably two or three questions. Let's see. I'll ask what type of experience you've had with the governmental affairs people; by that, I mean just a quick definition—something like, 'The governmental affairs department worked on a bill that helped our department,' and so on. Then I'll ask you to give me some of the details of what happened as you worked with the department. I'll probably also ask you to talk specifically about the people in governmental affairs—what they were like, how they worked, things like that. Then I'll probably also ask for your personal feelings about their department in general, why you feel that way, and how you feel they help the company as a whole."

"I'll never remember all that," the nervous one says.

"You don't have to," John reassures him. "I'll just ask you questions, and all you have to do is answer. Honestly, it really is simple if we just concentrate on talking and forget the rest of the stuff."

Thirty minutes later, John is sitting on a stool beside the studio camera, facing the nervous employee. Tape is rolling. "OK, Marty," John says. "Tell me about your experience with the governmental affairs department."

"Uh, well, they helped us. I mean, they sort of, I mean . . . oh, that sounds horrible. Can I start over?" he says.

John smiles. He knows that the best thing to do in this case is to just remain relaxed and wait Marty out. Soon Marty will begin to relax and converse, and John will get what he needs. "Sure, you can start over," he says, "anytime you want. Or, if you need to think about it, just take a minute. We've got plenty of time. By the way, let me ask you, Marty, do you really believe in this stuff you're going to say, or were you kind of told to do it?"

"No, I believe in it," Marty says, forgetting for a few seconds that the camera and lights are present. "They really are a hardworking bunch of people."

As he says this, John knows he has just gotten one very short but also very usable sound bite. "Well, for instance, what did they do that makes you say that?"

"They were just very . . . ah, dedicated. I mean, well, they worked long hours, for one thing. I know, because I didn't leave myself until late, and they were still there. But . . . ah . . ."

Again Marty loses his train of thought and becomes self-conscious, but he is making progress. He is starting to talk. "See," John says, "what you just said was fine. All you have to do is tell me about your experience."

Marty does start to make sense. Slowly, he becomes more relaxed, and eventually, John can tell he has got enough material for several usable pieces.

He releases Marty, and while he is getting the next person comfortable, Bill, a grip, and a gaffer quickly change the background of the set to make it look different. They slide in a flat, hang a decorative calendar on it, and place a plant in the frame on the other side.

To the camera, this person appears to be at a totally different location. To vary the shot even more, John moves around to the other side of the camera so that the person being interviewed will be looking in a different direction than Marty.

This person, Margaret, is more relaxed but too monotone and reserved. John knows that trying to get her to act would probably be a mistake, but he does try a simple form of method acting that has worked for him often during interviews like this. "That's great, Margaret," he says. "We can use a lot of what you've just said. Now I want you to do me a favor and be just a bit more forceful. Ever talk to, say, a neighbor or a relative and feel like you're not getting through when you really want to? Like you've got to lean forward in your chair and almost say, 'Hey, you. Listen up!' Ever done that?"

"Yes, I have," Margaret says.

"Well, consider me like one of those people. You've got to work just a bit harder to get my attention. It might not even hurt to think back to a specific occasion or person you had trouble with and make me that person. OK?"

"Sure. I'll try."

She does, and her performance is definitely better.

The next person is the same. He simply needs time to relax, think, and begin to feel at home in front of the camera.

John knew that the last thing these people needed was any suggestion of a Hollywood image on his part or any emphasis placed on the production aspect of what they were doing. He knows that to come off simply as a friend who is there to help and provide guidance is almost always a productive approach.

John has found that with camera-shy employees, this type of treatment and a good deal of patience almost always work. In this case, they do again.

By 5:00 P.M., the shoot is completely in the can, even including the impromptu interviews. John gathers the crew for a final meeting. After thanking them sincerely for a job well done and for their input and support, John releases them, saying he hopes to work with each of them again soon.

He then meets with Angela and reflects on the shoot for her. He tells her that he basically felt it went fine and that he is looking forward to meeting with the rough-cut editor the following week.

Angela is pleased. She and the client have spent a good deal of time on the shoot, and both have felt secure that John knows his stuff and works very well with actors and crew members.

Thus, everyone leaves the shoot with a feeling of confidence, and John with a feeling of deep reward and excitement. He is looking forward to the edit, for it is here that he sees all the previsualization he has done come together into a full program.

Postproduction

23

Three days after the shoot, John receives a call from Angela. She tells him that the off-line edit of his program will begin on Monday of the following week. She also mentions that the off-line editor is a woman named Gloria Tillance, who has done extensive work for her group.

"Good," John says. "I haven't worked with her, but I did meet her once when I was in for my last production there. She seemed like she'd be easy to work with."

"She will be," Angela says, "and she's a very good off-line editor to boot. So I'm sure she'll make your footage look very nice."

This comment about the off-line editor having an effect on the director's footage is accurate. John is well aware that the way the editor cuts together the footage will have a profound impact on both the aesthetic look of the program and its tone, pace, and content. It is because the off-line editor plays such a pivotal role in the development of any program that John will make it a point to sit with Gloria and explain his preferences at key points during this edit.

On the following Monday, John steps into an off-line edit bay in Angela's department and meets Gloria. He finds her looking over the master script, noting each of the buy takes and various other notes made during production. This is the script kept by Jenny during the entire shoot; a portion of it looks like the example in Figure 23.1.

Gloria also has a copy of the shot report kept by Jenny. Whereas the master script highlights each scene, noting what was good or bad about it, the shot report simply lists scene numbers and the videotape reel and time code location of each take (Figure 23.2).

The time code is a digital time "signature" recorded on each tape in a special "cue" track. It appears visible on the third element in Gloria's possession—the ¾-inch "window dubs" of all the footage recorded. The window dubs are work prints of the footage. Besides having the time code recorded on their cue track, they each have it "burned in" on a rectangle superimposed over the picture.

These two manifestations of time code will allow Gloria to find and edit precise segments of the footage and also to create the edit decisions list (EDL).

As we have discussed, the EDL will be used to execute the final or on-line edit of John's program. This is possible because the time code shown on the ¾-inch window dubs corresponds *exactly* to the same pieces

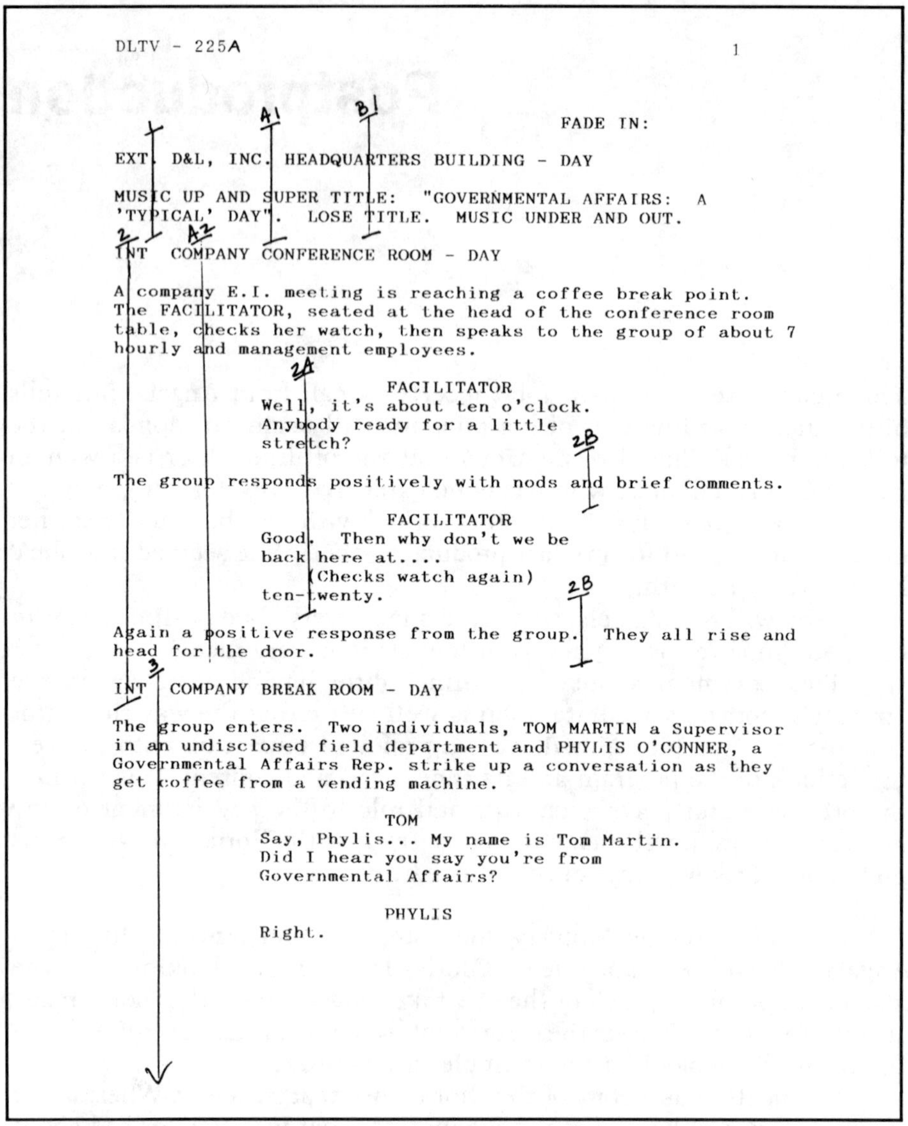

DLTV - 225A 1

FADE IN:

EXT. D&L, INC. HEADQUARTERS BUILDING - DAY

MUSIC UP AND SUPER TITLE: "GOVERNMENTAL AFFAIRS: A 'TYPICAL' DAY". LOSE TITLE. MUSIC UNDER AND OUT.

INT COMPANY CONFERENCE ROOM - DAY

A company E.I. meeting is reaching a coffee break point. The FACILITATOR, seated at the head of the conference room table, checks her watch, then speaks to the group of about 7 hourly and management employees.

FACILITATOR
Well, it's about ten o'clock. Anybody ready for a little stretch?

The group responds positively with nods and brief comments.

FACILITATOR
Good. Then why don't we be back here at....
(Checks watch again)
ten-twenty.

Again a positive response from the group. They all rise and head for the door.

INT COMPANY BREAK ROOM - DAY

The group enters. Two individuals, TOM MARTIN a Supervisor in an undisclosed field department and PHYLIS O'CONNER, a Governmental Affairs Rep. strike up a conversation as they get coffee from a vending machine.

TOM
Say, Phylis... My name is Tom Martin. Did I hear you say you're from Governmental Affairs?

PHYLIS
Right.

A

Figure 23.1 Master script pages (A and B).

of footage on the original material recorded. It is with this high-quality original material that the on-line or final edit of the program will be made.

Thus, John knows that the key aesthetic decisions will be made in the off-line edit—that is, the choice of what coverage to use and how it should be cut together. The final *polish* to this work—such elements as music, dissolves, wipes or DVE moves, and the insertion of titles—will be included in the on-line edit.

John sees that Gloria has also been reviewing some of the footage. He introduces himself and says, "Well, I thought we could look over the script and some of the footage. I'd like to show you some of what I had in mind with certain coverage."

"Fine," Gloria says, "Where shall we start?"

SCN	R#	Desc.
1	4	Ext. H.Q. W.S. lock off - pushes in to closer ① :45 & :45
A1	4	Alt. -Z.I. to Frn & Ent. 1. N.G.- wobble 2. too fast 3. GD ④
B1	4	Alt. Z.O. to W.S. - starts on front door 1. OK 2. too fast 3. N.G. - employee looks @ cam. 4. N.G. - clouds ⑤
2	11	Master -W.S. class in session. Facilitator releases employees & all exit 1. Fac. bobbled line · A/c running 2. Clumsy motion 3. OK 4. OK - mic. rustle between lines ⑤
A2	11	Alt. master - Dolly to reveal exit as employees leave 1. N.G. - ruff move 2. N.G. move 3. Keeper - end part best 4. N.G. - audio boom in shot 5. poor perf. 6. GD ⑦
2A		M.S. Facilitator -look off 1. OK 2. GD ③
2B		Audience reversals -W.S., M.S., 2 shots, singles - includes exit - 1st part ① Variety of focal lengths - editing choice

B

"Let's start by talking through the script," John says.

At this point, he moves through the script, providing Gloria with a brief overview of the type of coverage she will find on each scene and his preferences for how to use it. For instance, he says, "Scene 1 is an exterior of the headquarters building shot from a hill across the freeway. I recorded three versions of the shot—a wide lock-off with the freeway visible in the foreground, a closer lock-off that shows just the building, and a slow zoom-in. I'd like to try the slow zoom, since we'll super the title over it with some opening music."

As John says this, Gloria looks at her master script and shot report (Figures 23.1 and 23.2). She finds that the shot John is talking about is located on reel 4 and the time code noted is 04:02:31:00.

In this number sequence, the 4 indicates the hour and, also in this

VIDEOTAPE SHOT REPORT
FORM PR 2525

V- 225A	TITLE "G.A. Typical Day"	REEL NO. 4	DATE 2/21
PRODUCER A.S.	DIRECTOR S.M.	A.D. J.D.	

FOOTAGE IN	FOOTAGE OUT	SCENE NUMBER	TAKE NUMBER	DESCRIPTION
4:00		–	–	B & tone
4:01		1	(1)	ZI @ 4:48
4:02:31		A1	1	
			2	
			3	OK
			(4)	GD
4:05:06		B1	1	OK
			2	
			3	People look at cam. - NG
			4	
			(5)	Director likes end frame
4:09:13		9-1	1	
			2	
4:11			3	
			4	
			5	
			6	
			7	
4:16:20		9-3	1	Split scrn MS/MCU
			2	" "
			(3)	
			4	Full MS
			5	
			6	
			(7)	

Page ____ of ____

Figure 23.2 Videotape shot report.

case, the reel number of the footage. The 02 indicates the minute. Gloria is therefore aware that the shot is about 2 minutes in on reel 4. The 31 indicates the seconds, and the 00 is the frame number, of which there are 30 for every second of recorded video.

"OK," Gloria says, "according to my notes, there were actually two zooms also, a fast one and a slow one. I take it you want the slow one?"

"Right," John says, "It should run about 7 seconds to be able to accommodate the opening title, and I want to be sure it never runs out. We'll dissolve out of it to the interior shot while the zoom is still pushing in."

"I see," Gloria says. "And the interior we'll go to is the classroom?"

"Right," John says, "Scene 3. I think we bought take 11."

"Shall I do a split edit and start with some of the classroom audio under the exterior push-in—as a way to cue the dissolve?"

John hadn't thought of this, but he likes the idea.

"Yes," he says, "that sounds nice."

After saying this, however, he looks at his script and has second thoughts. In the opening classroom scene, the facilitator's lines are short. John's concern is that if his audio starts while the audience is on the exterior still seeing the zoom-in, by the time they get to the classroom shot, what's left of the facilitator's lines may be too brief to visually establish what's going on.

With this in mind, he qualifies his comment to the editor, saying, "You know, as I look at this, I'm afraid we might clip too much of this shot if we do an audio under split edit."

"If I use only the first sentence," Gloria responds, "I think you'll still have enough to establish the classroom."

"I think you're right," John says. "But let's keep it to the one sentence."

"Sure thing," Gloria says, making a brief note on her copy of the master script. "So that brings us into the classroom."

"Right," John says. "I shot that two ways. We recorded a wide master from the rear of the room, looking over the heads of the students, that has all the action. That would cut with a series of reaction shots we got from around in front. But I also got a dolly shot that starts at the rear and moves around, revealing the students and the door as the two principals leave through it."

"I'd go with the dolly if it works well," the editor says.

"I agree," John responds, "and I think the last take we got—4, I believe—worked really well."

"Good. Then I'll use it."

"OK, now that cuts with the first studio material. If I'm not mistaken, 8 was the buy on that scene."

"Shall we cue that one up and have a look?" Gloria asks.

"Sure," John says, as Gloria inserts a reel of tape into her playback machine.

The conversation goes on this way for several hours. When they part ways, John and Gloria have discussed nearly all parts of the script and footage, and John feels comfortable that Gloria will do a professional job. It has become obvious to him that she knows her stuff and also has good creative input for improving on what John has already done.

When John leaves Gloria, he meets briefly with Angela. "I think the off-line is in good hands," he says.

"Right," Angela agrees. "Just like the shoot was in good hands."

"Thanks," John says, "I enjoyed it, as always."

"And I enjoyed having you back again," Angela says. "I'm sure I'll be calling again soon."

"Anytime. Now, on this one, I'd like to drop in a few times to see how the cut is going, and if you want, I'm available to meet with you and the client for the rough-cut review."

"Let's see," Angela says, looking at her calendar. "The rough cut is due to be completed on Thursday. That's when I'll have a look at it and give Gloria my changes. There's one more day after that for any final revisions, and the client comes in on Monday. You don't have to come in for that one, John, if you don't want to, but you're certainly welcome. It's up to you."

"I'd like to," John says, "if you don't mind."

"Not at all."

"Good," John says. He realizes that Angela, like other producers he works for, will have the final say on how his program cuts together. He also realizes that she may want changes according to her personal preferences. Typically, the director is given "first cut" if time and budget allow, and the producer then has the editor make final rough-cut revisions.

John has found that meeting with the client and the producer offers a final opportunity to see how the footage has come together, and is usually a very positive way to end the job. It also leaves a professional reminder in the client's mind of John's abilities. This can often result in additional work, since many clients request certain directors for their projects.

John also knows, however, that such meetings are a time to be critical of a director's work. If he attends, he knows he must, first of all, not allow himself to become defensive and, second, never question Angela's judgments or decisions.

To become defensive is unprofessional, and to challenge any final decisions Angela has made on the show might be argumentative or disrespectful of her authority as the producer. To do so, John knows, would almost certainly mean that Angela would have second thoughts about bringing him back in for more work.

This, John knows, doesn't mean he cannot have input. It simply means he must be careful in such meetings to respect Angela's wishes and authority before making any suggestions.

"Well, then," John says, getting to his feet, "I guess I'll be going. I'll check back, and I'm sure you'll have no problems with the show."

"I'm sure you're right," Angela replies, getting up from her seat and shaking John's hand.

John returns on Tuesday. Gloria is roughly one-third done with the show, and it looks fine. His only suggestion is that the montage, scene 5, have more edits, for a faster pace and more moving shots.

Gloria has cut the scene using three static shots. John says, "The lock-off shots look fine, but the intensity of the work these people do when they're under pressure could be brought out more, I think, by picking up the pace and using the moving hand-held shots."

"Actually," Gloria says, "I thought about that, but when I looked at the shots, I felt they might distract. Some of them were pretty shaky."

John recalls that she is right. Some of the shots were fairly shaky, but there were others that were also steady. He is firm in his belief that the montage needs more tension and a faster pace. "If you want," he says, "I'll go through some of the material with you. I remember I had Jenny circle the best of it for each shot."

"Do you really think we need to rework it?" Gloria asks. She is not thrilled about the idea, because it means reediting the scene that follows it as well.

"Yes, I do," John says. "The show looks really nice, Gloria, but one of the main things we were trying to accomplish with this was to make the point that these people have to work under extreme pressure at times and that they're dedicated and hardworking. I think if we go with the mild version of these montages, we'll lose some of that impact."

"OK," Gloria concedes. She, like John is a professional, and, as such, she realizes that what he is saying is correct.

On Thursday, John sees and likes the complete rough cut. He thanks Gloria for her hard work and professional input and says he hopes to work with her again.

On Friday, he drops in and sees Angela's changes, which are minor. Although he doesn't agree with them completely, he feels they are insignificant enough to negate the value of any arguing.

On Monday, John attends the rough-cut meeting, and the client is pleased. He too has several changes that, in John's estimation, are minor and not worth arguing over.

Following the meeting, John and Angela have lunch.

"Can I get you back in on a project I've got coming up next month?" Angela asks John.

"Sure thing." John says. "I got a call from Abbot and Sons yesterday. They've got a couple of short segments they want me to shoot for their in-house newsmagazine show, but that should wind up in a week or two. After that, I'm open."

"Great," Angela says. "Then drop in and have a look at the final edit on this show. It's going to look great. And maybe by then, this other project will be ready for you to have a look at."

"Sounds good," John says, taking a first sip of his soup and beginning to think about how he might shoot the news segments for Abbot and Sons . . .

PART IV

A Career

Starting Out 24

Looking Back

Years ago, as a young man preparing to enter the job market, I remember being urged to set career goals and objectives. At that time in my life, such talk meant little to me. Goals and objectives were "academic." I was confident that if I simply knocked on a few doors, opportunities would present themselves immediately and things would fall neatly into place. Having now traversed a full corporate career, when I look back, I realize that I was not quite as special in those early days as I'd led myself to believe.

I *have* achieved a good deal in my career, but the majority of it has taken a great deal of time and work, and it has materialized only after a good many hit-and-miss efforts. In addition, most of my achievements have come only after I began to focus on my future and develop a very strong sense of exactly what I wanted to accomplish—in other words, goals and objectives.

With this in mind, my first suggestion to those readers who have *decided* on a career directing corporate or other types of media is this:

1. Get the best possible education, both in college and through reading, and by personally meeting with and learning from experienced crew people, producers, clients, and directors.
2. Consider the achievement of a respected directing career in whatever medium you choose as your *long-term* career goal.
3. Analyze and list out the smaller achievements you must accomplish in order to reach your long-term goal. Each of these smaller accomplishments should be considered your *short-term* objectives. Read and ponder them continually.
4. Once this career plan is in place, get to work on it and let nothing deter you from accomplishing it—not even the painful process of rejections that is an inherent part of any creative aspiration.

If your long-term career goal is to direct corporate productions or even feature films, your next question might be this: Just what short-term objectives should I focus on? While it is impossible to map out exactly how any individual might reach top-notch directing status, there are several fairly standard career stepping-stones.

1. **Learn and experience all aspects of the production process.** This means doing at least some work as a production assistant, grip, gaffer, floor manager, camera operator, editor, lighting director, sound assistant, and assistant director—and, preferably, as a writer and actor as well.

   As you become involved in each of these activities, make it your personal business to learn everything you possibly can about them—especially how they relate to the director's job.
2. **Don't get stereotyped by producers.** It is very typical for college interns or graduating students to begin working in production facilities as grips and production assistants. Some newcomers, however, tend to stay in these positions and, after a period of time, become thought of only in that regard.

   If your plan is to someday direct, you must work in various crew positions, but *only* long enough to become proficient. You must then seek and move on to new career challenges in keeping with your short- and long-term goals.

   This can sometimes mean turning down work. But it can also mean sitting down with the producer you work for and making it clear that your long-term career goal is to direct and that you would *jump* at any additional challenge he might provide that could assist you in reaching this goal. If you are a grip, ask to gaff. If you are a gaffer, ask to light and shoot. If you are a DP, ask to direct some simple voice-overs or cover footage.
3. **Sharpen your aesthetic sensitivities continually.** As we have seen, the critical aspects of a director's work spring from observations, sensitivities, and judgments based on personal preference. In order to be able to execute these judgments with continued success, you must know what works. Knowing what works is really an extension of your aesthetic sensitivity to the human and technical elements of your project.

In addition to the basic steps just mentioned, there are a number of additional actions you can take to help develop your directing background.

Find and Work for a Mentor

If you find yourself in a position of working for or making the acquaintance of a director who you respect, learn everything you can from him. Be very direct in stating your long-term career goals, and ask if perhaps he will consent to passing on some of his knowledge to you in exchange for your support and hard work as an assistant. No amount of reading or classroom work can match the value of the lessons you may learn on the job from a seasoned director who is willing to pass on his personal "secrets" to you.

Study the Corporate World

As we have discussed, there is also the business side of a corporate director's work. This means knowing and being able to operate comfortably in the corporate world. You can prepare yourself for this eventuality by learning as much as you can about how corporations in your area operate.

Visit companies and talk with public affairs representatives. Ask if they develop corporate media programs and, if so, how you might be put into contact with that department. Also learn the rules of business protocol and etiquette. Learn how to contact businesspeople, and then use those contacts as liaisons to other departments. Learn about management levels and what business rules apply to lower-, middle-, and upper-level managers. Learn to be comfortable with top executives, and understand their concerns. With this sense of business savvy under your belt, you will be prepared to work with and gain the trust of corporate decision makers who will have a profound effect on your work.

Direct

When asked how to break into directing, one director said simply, "By directing." Although this may sound like an oversimplification, it has a great deal of truth to it.

You may not have a production crew at your disposal and top-notch equipment, but why not actually direct your own pieces with the resources at hand? Use your home camcorder and whatever other equipment you can round up. Create dramatic scenes using friends or relatives as actors, or start by shooting documentary-type footage—perhaps an informational piece on some aspect of your city. Make simple voice-over programs using VCRs to dub music and narration under pictures you've edited together.

Granted, these productions won't land you any awards, and, in fact, they may not even be anything you'd care to let anyone see. But they will put you in charge behind the camera and expose you to the same types of judgments a director must make continually.

Why did a certain kind of framing work well? Why did another seem flat—was it lighting, lack of depth, an unbalanced frame? Did one of your "actors" sound comical when she was supposed to have been dramatic? If so, what *exactly* was it about her performance that made the difference—her delivery of the lines, her facial expressions, her movements in front of the camera? After watching your piece, do you think the subject was communicated clearly? If so, why? If not, why not—faulty writing? faulty direction? a combination of the two? By going through this process, you will begin to sense which elements work and which don't and why.

Study Acting and Writing

Many directors are also scriptwriters and actors. Perhaps this is because the three professions are so closely linked and require many of the same sensitivities and creative skills. In addition, as we have seen, two of the most important aspects of a director's work are communicating with actors and analyzing scripts.

Having writing and acting skills, then, not only allows the director to communicate better with those with whom he works most closely but also offers constructive perspectives from the other side of the fence, so to speak. This continually sharpens a director's perception and communication abilities.

As an aspiring director, you might volunteer to act in local theater groups, student films, or the personal work of other directors who perhaps are your friends or mentors. These same avenues offer possible outlets for your writing skills.

Study Art and Photography

A good foundation for the visual aesthetic skills used by the director can be obtained by studying art and still photography. A close similarity exists between the proper frame size and balance in many paintings and still photos and the moving frames of video or film productions.

Each requires a sense of depth, lighting, and perspective. Each also involves the use of planes of focus and various placements of people and elements in the frame. Just as with a painting or photograph, video and film production is often enhanced by a refreshing choice of angles or unique frame composition.

One way this similarity is evidenced is in many high-budget commercials seen on TV. Watch them carefully and you will notice that some are like a series of perfectly composed paintings or photos with moving instead of still elements in the frame.

Internships

If you are able to live on little or no money for a number of months, an internship at a corporate television production facility may be invaluable. Just as with any craft, only so much can be learned from reading and even hands-on training. At some point, you must listen to actual clients discuss concerns with a director. You must watch real producers make decisions and handle the various aspects of managing corporate productions. You

Case Study 10 *Multicamera Madness*

One director I know, who was experienced at shooting single-camera productions, took on a multicamera shoot and got himself in over his head.

He was very proficient at lining up shots individually and working out focal lengths and continuity for matching coverage. When he found himself having to spontaneously switch among three cameras, however, he became very nervous, clumsy, and disoriented.

That's because, besides the usual aesthetics, multicamera directing requires a sense of timing and rhythm that comes very naturally to some directors but not to others.

In this case, the director ended up shooting the majority of his show with a single camera and letting the other two (including the operators) sit idle.

must carefully consider a director's actions on the set or in preproduction or postproduction and ask yourself what you would do in her position. Moreover, internships have an additional value: the introductions you will make and the opportunities to leave impressions that may later result in work for hire.

Many corporate facilities seek student interns during the summer months. If a formal internship is not available through your college but you have the luxury of some form of temporary monetary support, simply approach the producers in your area and volunteer to serve as a free assistant of some sort.

Your eagerness alone will impress them favorably, not to mention the budget break a free assistant might provide on their next production. Remember, however, that if a producer takes you up on the offer, you must be ready to start with very menial tasks, such as filing, sending out tapes, or making calls. Simply do an excellent job, maintain your enthusiasm and willingness to help, and wait patiently for your own break to materialize.

Community Access Cable

Most local cable channels must, by law, allocate a certain amount of time and resources for the production of what is called community access programming. This simply means programming created by and used to help those in the local community. Toward this end, most cable channels also give simple classes in production skills and allow equipment to be borrowed for use in making community access programs.

These programs are rarely stunning or fancy in terms of sound and image quality or production techniques. They are, however, one more way to create opportunities and gain on-the-job experience. They also provide a chance to meet local people in the business and allow for public broadcast of your work.

In Closing . . .

In the final analysis, building a career as a director of films or videos can be a very difficult road. Rejection is commonplace and the competition is fierce for the few top jobs that exist.

For those who tough it out, however, and finally succeed, the rewards are great. I hope you are one of those future success stories, and I sincerely hope this book has provided some amount of guidance toward the achievement of that goal.

With that, let me simply say, "Good luck, good directing!"

APPENDIX

1

Needs Assessment

NEEDS ASSESSMENT

PRODUCTION NUMBER	225A
DATE	JUNE 1, 1992
WORKING TITLE	GOVERNMENTAL AFFAIRS: A "TYPICAL" DAY
SUBJECT	THE GOVERNMENTAL AFFAIRS FUNCTION IN D&L, INC.
CLIENT	DARREN WHITESALL, ADMINISTRATOR, GOVERNMENTAL AFFAIRS, 3170/500, 555–6801
OTHER APPROVALS	HAL MORTON, DIRECTOR, GOVERNMENTAL AFFAIRS, 3170/500, 555–6540
	KEITH MANSUR, VICE-PRESIDENT, REVENUE REQUIREMENTS, 3400/500, 555–7130
CONTENT EXPERTS	DARREN WHITESALL (AS ABOVE)
FYI COPIES	TO BE DETERMINED
DEADLINE	SEPTEMBER 15, 1992

COST/BENEFITS/IMPACT

The cost of this program will be approximately $40,000. Calculating its value to D&L, Inc. is difficult in monetary terms.

ASSESSMENT
Page 2.

We do know that approximately 600 legislative bills were passed in 1990 that had some effect on D&L. Of those, approximately 300 had direct impact.

It stands to reason that employee understanding and support in this area should make the governmental affairs function a more effective one in shaping a healthy future for our company.

In addition, to distribute a message of the type proposed in this document without the use of videotape would require live presenters. This would mean visits to numerous D&L locations over a period of months, at a cost that would far exceed the $40,000 production cost mentioned earlier.

PROBLEM

Many D&L employees do not understand what the governmental affairs function is, how it operates, and the fact that it has strong executive support.

BACKGROUND

To date, very little has been done to educate employees on the workings of the governmental affairs department. Videotapes have been produced on the Good Government Club, but this is an independent organization with a different purpose.

RELATIONSHIP TO COMPANY GOALS

D&L's goals center on providing top-quality service at affordable prices. Legislature is continually passing through the state capitol that can affect our ability to do this. A strong governmental affairs department, then, with broad employee support, is in a much better position to help the company achieve its cost/quality goals.

AUDIENCE

Primary: Middle- and lower-level managers in D&L, Inc.
Size: Approximately 5000

ASSESSMENT
Page 3.

AUDIENCE DEMOGRAPHICS

Most audience members have college degrees and hold positions of responsibility in D&L. Approximately 65% are males who have considerable seniority with the company.

AUDIENCE INTEREST/NEED

Once viewers realize the amount of governmental activity D&L is involved in and the impact it can have on their individual futures, their interest in the subject should be considerable. Prior to this realization, the interest level should be fairly low. For this reason, a teaser or grabber of some sort will be used to make this point in the beginning of the program.

AUDIENCE KNOWLEDGE/EXPERIENCE

Many employees are surprisingly uninformed about the workings of our government in general and its potential impact on D&L. While this will not be an instructional program, it will impart certain basic information in these areas.

Most employees also have little knowledge of what goes on in the governmental affairs department and how its function might affect their futures.

AUDIENCE ATTITUDES

Many audience members probably have general areas of distrust about state and federal government and a "so what" attitude about its relationship to D&L. This is due to the general lack of knowledge mentioned earlier.

Also contributing to this negative attitude is a lack of knowledge about what governmental affairs representatives do on a day-to-day basis and the misconception that their work is somehow plush and easy.

This program will attempt to change these attitudes primarily by revealing the hard work and often very long hours put in by the governmental affairs employees in our company.

ASSESSMENT
Page 4.

OBJECTIVES

Primary:
Having viewed the proposed videotape, audience members will be able to state

1. The role of Governmental Affairs in the D&L organization: that of a political watchdog and key protector of D&L's legislative interests
2. One way the Governmental Affairs department helps secure the future of D&L Californians: by monitoring state legislature and acting on those bills which may have an impact on D&L
3. That the Governmental Affairs department in D&L has full CEO and executive support

Secondary:
Having viewed the proposed program, audience members will be supportive of D&L's Governmental Affairs department and the hard work done by its representatives.

DISTRIBUTION

Distribution of this program is still being discussed.

UTILIZATION

Utilization will most likely be in conference rooms throughout the company. Final plans for utilization, however, will be determined with distribution plans.

EVALUATION

A formal evaluation of this program will take place upon completion.

APPENDIX

Treatment 2

❑ ❑ ❑ ❑ ❑

CHARACTER ANALYSIS AND SAMPLE TREATMENT

"AFTER HOURS"
A Corporate Family Situation Comedy/Drama

SERIES SYNOPSIS

"After Hours" is a continuing situation-comedy series that explores various important issues affecting hourly and management employees in CommRac. The series attempts to examine issues of great concern to all CommRac employees and resolve each issue it focuses on with a solution supportive of the company's vision and values.

"After Hours" takes place in the middle-class home of the Martinez family. Each family member has a strong sense of personal values and definite convictions about such issues as work ethics, safety, drug/alcohol abuse, labor relations, management responsibilities, and craft responsibilities. It is these strong convictions that provide the often conflicting opinions and dynamic examinations of the issues explored in "After Hours."

THE CHARACTERS

The main characters of "After Hours" include the father, HENRY MARTINEZ; his wife, ANGELA; their son, ROBERT, who still lives at home; and their daughter, PENNY, who is married but visits often. Penny's husband, ANDY, is also a continuing role, as is TRACI, Robert's girlfriend. Various other friends and neighbors also drop in occasionally.

"After Hours" 2

HENRY

Henry Martinez is 62 years old. He is a newly retired, "heavy gang" (outside plant construction) foreman—not a supervisor, Henry would be quick to tell you, but a *foreman*. He " . . . ran a crew of great guys, back when the phone company was *really* the phone company—not some high-tech sales outfit!"

Although Henry's opinions of today's telecommunications industry aren't the most positive, he is nonetheless a good man with strong values and a great deal of "ground-level" wisdom about telephony—wisdom that at certain moments makes more sense than all the computer programs and marketing projections in the industry. Wisdom or not, however, Henry's entrenched, male chauvinistic attitudes often put him at odds with other family members on issues being discussed.

ANGELA

Henry's wife, Angela, is 58 years old. She is a housewife with time on her hands, since the kids have grown up. Angela spends a good deal of her energy keeping both Henry and Robert in line. Though she may not be the image of today's modern woman on the surface, she's no pushover when it comes to handling men and exerting her influence over the family and the issues and problems it faces. Angela is also the unlikely visionary of the Martinez family. Although all other members of the family are more educated and deal with the business world continually on one level or another, each is influenced by various personality traits. Angela, on the other hand, has the uncanny ability of being able to see things as they really are, regardless of the emotional content of the issue. This seeming paradox becomes apparent shortly after we get to know the family and its members.

ROBERT

Robert is the 22-year-old son who lives at home. He is a supply and transportation craft employee and a union steward. Robert is as pro-labor as his father is pro-old days. He believes strongly that his company has an inherent commitment to its employees in terms of fair

"After Hours" 3

ROBERT (cont.)

wages, working conditions, job security, and so on. His personal opinion on how well the company is living up to that commitment causes him mixed feelings. While he feels CommRac is basically a good company to work for, he's angered at recent efforts to trim down the work force. Along with this anger, Robert has a general disrespect for upper management. In a nutshell, he's not totally convinced the "guys in the ivory towers" know what they're doing or that they're giving the people on the front lines a fair shake.

PENNY

Penny is 24. She is Robert's big sister. Penny dresses very "in," she's quick to speak out on the issues, and her point of view is always pro-management. She works for a large computer company, INB Information Systems, as a senior account executive (a salesperson). Penny knows labor ratios, stock market implications, market shares, business trends, and so on.

Penny's primary shortcoming is the exact opposite of her younger brother's. She's a little too sympathetic to the white-collar side on the management–craft issue. In her opinion, unions are wasteful organizations run by crooked, money-hungry gangsters who want to get paid top dollar but don't want to work. She's convinced that unions have no place at all in American businesses.

It goes without saying that Penny and little brother Robert go "round and round" frequently on these issues and don't often reach common ground.

ANDY

Andy is Penny's husband and a professor of psychology. He is a mild-mannered yuppie who angers both Henry and Robert because he is a "contemporary" male—that is, he does dishes, irons clothes, helps cook, and so on. He also gets "put in his place" often by wife Penny and thus sometimes comes off as the comical whipping post during the frequent Martinez family "discussions."

Andy is from a WASP background and isn't quite able to rationalize the "fiery" Latino temperament of both his wife and the Martinez family in general.

"After Hours" 4

ANDY (cont.)

Regardless of these differences, he and Penny basically love each other very much and Andy respects everyone in the Martinez family.

TRACI

Robert's 21-year-old girlfriend, Traci, is in college. Her major is business, and she's somewhat of a rebel. She rides a motorcycle and likes to admire passing "hunks"; her favorite pastimes are bass fishing and basketball.

Traci and Robert tend to mix it up a bit on occasion, since he is the jealous and somewhat dominant type. Henry, however, is also a fisherman and basketball fan, and so he and Traci hit it off quite well.

CHARACTER/SITUATION SUMMARY

Using a Hispanic family as a basis and the character types just mentioned should produce a believable mix of opinions and ample motivation for conflict on various issues. In addition, the issue of minority representation is resolved, and the level of audience identification should be high with this group of contemporary yet traditional characters. Audience trust and respect for the characters should be high also. In other words, the audience may not agree with every opinion put forward, but they will respect the fact that it's a sincere opinion and well intended.

A SAMPLE STORY TREATMENT

"AFTER HOURS . . .
Strictly Business?"

SYNOPSIS

Robert's best friend, Carl, a carpet installer who has worked for a large company for 10 years, is suddenly laid off. When he drops in at the Martinez house to share the bad news with Robert, the entire family becomes involved in a heated discussion. The issue? Whether or not layoffs and recent trends in downsizing are fair to workers and a justified business option.

"After Hours" 5

THE PROGRAM

It is Saturday afternoon at the Martinez residence.

We fade in on Robert and Carl having a depressing conversation in the Martinezes' garage. They are discussing Carl's recent layoff, and questioning what will happen now to Carl himself, his young wife, and their 2-year-old daughter. Robert is saying that many large companies are grossly mismanaged these days and that the front line workers end up paying for it with their jobs.

Henry Martinez enters, at first oblivious to what is going on. He is carrying a fishing pole and various freshwater fishing accessories—net, tackle box, and so on.

As he passes the two young men and hangs his pole on a rack, he mentions to the boys that he may have to cancel his annual, week-long fishing trip. When the boys ask why, Henry tells them that things are a bit tight this year and the money might have to be spent on a new car—that is, if the verdict is bad from the mechanic currently checking out their present car.

When Carl hears this, he mentions that at least the Martinezes can think about getting a car. In his (Carl's) situation, he may have to think about trading in his car for a bike!

The boys now tell Henry about Carl's predicament. Henry immediately starts remembering and, of course, verbalizing about the old days when a job meant security and commitment and bosses had the dignity to at least give an employee some notice before firing him.

From the house, Angela now calls to Henry to come and help start the potatoes, and Henry asks Carl to stay for lunch. Carl mentions that his wife and daughter are visiting a sister and accepts.

Later, Carl, Henry, and Robert are in the living room, still discussing the layoff. Penny and Andy now enter. They have come for lunch also.

As soon as Penny hears of Carl's layoff, she begins to justify the management side of the issue, much to the ire of Robert, who, of course, sides with his friend and the position of labor in general.

As the issue is getting heated, Angela comes in and quietly asks that the conversation stop and everyone come to lunch. No one seems to hear her, however, and so she tries again. Still no response. Finally, she walks into the center of the group. In one fell swoop, she twists Robert's ear, pulls Carl's baseball cap down over his eyes, kicks Henry in the shin, and, when Penny is about to say something else, sticks a finger in her face. She then proclaims simply, "Lunch . . . now . . . or not at all."

The group exits to lunch. It is now, at the dinner table, when the real issues are explored. During the conversation, CommRac's current streamlining trend is mentioned, as is that of Penny's company, INB. No final judgments are made, but all opinions are expressed. They range from Robert's belief that layoffs are a sign of mismanagement and should never happen to Penny's position that management's ultimate responsibility is to keep a company solvent and, as economics change, layoffs have to be considered a part of corporate life. Henry, of course, gets in his two cents—simply that he is glad to be retired in this cutthroat day and age.

Angela remains quiet through most of the discussion but finally makes an interesting point. She says that what appear to be setbacks can often be opportunities in disguise. The real issue, she says, is not whether or not the layoff is right or wrong, since it is now old news, but rather what Carl decides to make of it.

Carl isn't really buying this. His general mood is just plain down and negative. Just as he is delivering this response, the phone rings. It is Carl's wife. Although we don't hear her end of the conversation, it becomes clear to us that she is telling Carl something positive.

When he hangs up, Carl relays the word to the Martinezes: He has had a call from E.G. Carpets, a company roughly five times as large as Carpet Universe, which just laid him off. The personnel people at E.G. Carpets got word from Carl's former boss that he was a

"After Hours" 7

great worker and might be in the market for a job. Coincidentally, the company is bringing on a few new people and wouldn't mind raiding the talent pool of its competitor. Would Carl be interested in interviewing?

Carl turns to Angela and, with a word or glance, acknowledges that he now understands what she was saying just before the call came in—that setbacks can sometimes turn into opportunities.

Just then, the phone rings again. Henry answers. It is his mechanic and the news is good. The Martinez car isn't in the greatest shape, but since it's a collector's item, the mechanic has a friend who will pay top dollar for it. This means there may be enough money to both buy a new car and take the fishing trip!

Henry turns to Angela and poses this question to her, with a "Can I please go, sweetheart?" glance. Angela turns to the group. "The problem," she says, "is that some opportunities mean I have to stand at the sink and clean a week's worth of bass and trout!"

As this gets a resounding group laugh, we FREEZE FRAME on the scene and fade out.

APPENDIX

3 Script

PROJECT # 225A

GOVERNMENTAL AFFAIRS: A "TYPICAL" DAY

FIRST DRAFT: 7/15/92
SECOND DRAFT: 7/29/92
THIRD DRAFT: 8/12/92
FOURTH DRAFT: 9/7/92

DLTV - 225A 1

FADE-IN

EXT D&L, INC. HEADQUARTERS BUILDING—DAY
MUSIC UP AND SUPER TITLE: "GOVERNMENTAL AFFAIRS: A 'TYPICAL' DAY." LOSE TITLE. MUSIC UNDER AND OUT.

INT COMPANY CONFERENCE ROOM—DAY
A company employee involvement meeting is reaching a coffee break point. The FACILITATOR, seated at the head of the conference room table, checks her watch and then speaks to the group of about seven hourly and management employees.

FACILITATOR
Well, it's about 10:00. Anybody ready for a little stretch?

The group responds positively with nods and brief comments.

FACILITATOR
Good. Then why don't we be back here at . . .
(Checks watch again)
10:20.

Again the group gives a positive response. They all rise and head for the door.

INT COMPANY BREAK ROOM—DAY
The group enters. Two individuals, TOM MARTIN, a supervisor in an undisclosed field department, and PHYLIS O'CONNER, a governmental affairs rep, strike up a conversation as they get coffee from a vending machine.

TOM
Say, Phylis, my name is Tom Martin. Did I hear you say you're from governmental affairs?

PHYLIS
Right.

TOM
Listen, I'm in the process of looking for a transfer. I've about had it with my job. Could I maybe pick your brain a little?

Phylis gets her cup and moves to a nearby table.

PHYLIS
Sure.

DLTV - 225A 2

TOM
Great.

Tom gets his cup and moves to the table. He takes a seat.

TOM
I'll tell you, Phylis, this field-work rat race is really crazy—foul-ups, deadlines, customer problems. I'd like to get into something a little more . . .
(Thinks)
I guess controlled and relaxed are the words.

Phylis shoots a quick glance at Tom after the word relaxed.

PHYLIS
I see.

TOM
So tell me about governmental affairs. What exactly do you do locked away up there in the, ah . . .
(Smiles)
H.Q. building?

Again Phylis glances at Tom. Her expression tells us she's not happy with the attitude reflected in Tom's comment. She does answer, however.

PHYLIS
Well, basically, Tom, there are a lot of things happening in the state government every day that can have a big effect on the company.

TOM
For instance?

PHYLIS
For instance, last year there were 242 bills introduced that had a direct impact on D&L, Inc. in California. It's our job to keep a close eye on those bills. And we go out and get opinions and input from the departments in the company they can affect. From that, we develop what we call a company position, and we make sure it's heard in the capitol.

DLTV - 225A 3

Tom, it seems, is looking only for the ease and glamour. He smiles.

TOM

I see. So you fly to Sacramento a lot and wine and dine the politicians.

PHYLIS

Well, we've got some people in Sacramento, true. And we all attend certain functions. But I don't think "wine and dine" is really—

TOM

Schmoozing with the big guys? That more like it?

PHYLIS

Well, I mean, we—

TOM

Give me an example of some bill or something you guys might work on. Or what might happen in a typical day.

PHYLIS thinks for a moment. Suddenly, an appropriate example occurs to her.

PHYLIS

Well, actually, there was a piece of legislation introduced last month that was pretty typical. It started out as a bill we generally supported.

TOM

What was it about?

PHYLIS

The details are a little complicated, but basically it dealt with how we handle easements and boundaries . . .

As PHYLIS remembers, we . . .

DISSOLVE

FLASHBACK—INT GOVERNMENTAL AFFAIRS OFFICE—DAY

It is last month. The department DIRECTOR is just approaching Phylis's desk with a copy of the easements bill in hand. He drops it in front of Phylis.

DLTV - 225A 4

DIRECTOR
As if you need something else to do.

PHYLIS
(Quickly checks papers)
What have we got?

DIRECTOR
Deals with billing on easements and boundaries. Sounds like the author is thinking along our lines, but get on the phone to the field and the legal department and let's put a position paper together to send up to the hill.

PHYLIS
You got it.

DISSOLVE

MONTAGE
A series of shots depicting Phylis's activity in putting together a position paper on the bill in question. SHOTS INCLUDE Phylis on the phone to people in the plant construction department; visits to the legal department; a visit to the field, which includes a meeting on the job with a CONSTRUCTION SUPERINTENDENT; phone conversations with reps in Sacramento; and so on.

The LAST SHOT is Phylis at her desk with the position paper in hand, talking to the rep in Sacramento.

PHYLIS
I've been over it pretty well down here, Scott. We generally support everything in it. So it's coming up to you tonight. The ball is now in your court, my good man.

SACRAMENTO REP (VO)
Sounds like this one should be no problem. But I'll definitely spread the word in the right places. I've got a ground breaking tomorrow afternoon with some key people showing up—influential friends. I'll be sure they're on our side.

DISSOLVE

DLTV - 225A 5

PRESENT—INT COMPANY BREAK ROOM—DAY
Phylis and Tom are still sipping coffee. Phylis has just conveyed the flashback we've just seen. As usual, Tom has absorbed very little.

TOM
(With typical overconfidence)
Sounds pretty elementary. A few calls and field visits. I could handle that.

Phylis sees that she's not getting through to Tom. She decides to be frank.

PHYLIS
You know, Tom, I'm not sure you're reading me right on this. I mean, I enjoy my job a lot. It's rewarding, it carries a lot of responsibilities, and sometimes it can be exciting. But it's also serious business. I mean, it's plain old hard work. And some of the things we deal with can really affect this company in a bad way unless we—

Tom still hasn't really heard.

TOM
Like how?

Again Phylis thinks for a second.

PHYLIS
Well, take that same bill. Like I said, it started out to be no problem. But sometimes things change . . .

Phylis remembers . . .

DISSOLVE

FLASHBACK—INT GOVERNMENTAL AFFAIRS OFFICE—DAY
Trouble. Another REP comes past Phylis's desk, moving fast.

REP
Meeting. Right now. In the conference room.

Phylis checks her watch, grabs paper and a pencil, gets up, and exits quickly.

DLTV - 225A 6

INT CONFERENCE ROOM—DAY
The meeting is taking place around a large conference table. In attendance are the director seen earlier and FOUR governmental affairs REPS. Phylis moves into her seat and looks at the director.

PHYLIS
Trouble?

DIRECTOR
Big trouble. Remember that boundaries and easements bill?

PHYLIS
Yeah. It's all taken care of.

DIRECTOR
It was taken care of. It's up for a vote on Monday, and they've slipped in an amendment at the last minute.

PHYLIS
What did they do to it?

DIRECTOR
They went 180 degrees. If we don't get a "no" vote, we think it's gonna cost jobs. We need to get back to the people in the field and legal and come up with a new position. And by my watch we've got about . . .
(Checks watch)
. . . 6 hours to do it.
Anne, you get on the phone to legal right away. Phylis, you get with Dale up on the hill and start to work on the new paper. Andy, call the construction directors. Phylis has a list. Laura, get the fax machine fired up and be ready to type and send. I need to call Mr. Crain, and . . .
(Pauses, smiles)
order pizza.

REP
(Chuckles)
No anchovies, please.

DIRECTOR
Let's go.

DLTV - 225A 7

Everyone is up in a flash and headed for the door.

DISSOLVE

MONTAGE

A series of SHOTS depicting the rework involved in doing the new position paper—calls, typing, visits, and so on. Finally, we see Phylis at a PC inputting the new paper with the phone to her ear at the same time. We see, in a SPLIT SCREEN, that the person she is talking to is <u>the same construction superintendent</u> she talked to earlier. A coke and a half-eaten piece of pizza lay by her phone.

DISSOLVE

PRESENT—INT COMPANY BREAK ROOM—DAY

The memory fades, and Phylis returns to the present.

PHYLIS

That one got pretty hectic.

TOM

Sounds like a little overkill to me. I mean, calling the company president?

Phylis is now getting plain irritated.

PHYLIS

Look, Tom—

TOM

I know. I know. Jobs at stake, right? You know, the trouble is, I hear that kinda stuff all the time—"This'll cost jobs; that'll cost jobs." And, you know, I'm still here, man! Now, tell me the truth. I mean, really. If you hadn't redone all that work and stayed late and all that, do you really think anybody would have lost their job?

ON PHYLIS—She glares at Tom and remembers . . .

DISSOLVE

FLASHBACK—INT DIRECTOR'S OFFICE—DAY

It's Monday morning. The governmental affairs group from the meeting on Friday is gathered in the director's office. The director is seated behind his desk. He has a memo in hand. He looks up at the "troops" and pauses. After we see a series of REACTION SHOTS that tell us everyone is wondering what he is about to say, he begins.

DIRECTOR
This is from Mr. Crain.
(Begins reading)
Dell, a vote was taken on the boundaries bill you and your staff worked on over the weekend. You'll be happy to know the bill was . . .
(Pauses, looks up)
soundly defeated. . . .

A cheer from the group.

DIRECTOR (CONT.)
A new version without the amendment is now in the works. I'd like to pass on my thanks to you and your staff for a job well done under fire, and let you know that our research shows you've just helped save approximately 375 jobs. While most of those people will probably never be aware of what you folks have done, I'm well aware and very proud. . . .
(Pause)
Keep up the good work. Martin Crain. . . .

Proud smiles and other reactions from the group.

DISSOLVE

PRESENT—INT COMPANY BREAK ROOM—DAY

Phylis is back to the present. It's obvious she's proud of that memory.

PHYLIS
Yes, Tom. I really do think that if we hadn't done all that, some people really might have lost their jobs. In fact, one of the people I consulted with on the bill was a construction superintendent who probably would have lost half his people.

Tom still doesn't believe it. He chuckles and shakes his head at Phylis's dedication. Phylis now realizes she won't be able to get through to Tom. She takes a last sip of her coffee.

PHYLIS (CONT.)
Listen, I've gotta go, Tom. It was nice talking with you.

DLTV - 225A 9

Phylis gets to her feet.

TOM

Yeah, I've gotta go too. Gotta make a quick call to my boss before we go back to the meeting.

Both people exit.

DISSOLVE

INT HALLWAY—DAY

Tom is on the phone. He is talking to his boss.

TOM

Hi, Gary. It's me. I just wanted to let you know I expect to be back in around 3:00. How do things look?

As we go to a SPLIT SCREEN, we see that Tom's boss is the same construction superintendent Phylis had talked to several times during the work she did on the easements and boundaries bill. In other words, Tom's job was probably one that Phylis helped save.

GARY

Everything's moving along. Get back in as soon as you can, though—we've got a crew up north that needs help. . . .

MUSIC UP. We go to a CLOSE-UP of Tom and FREEZE FRAME. MUSIC OUT AND . . .

DISSOLVE

INT MARTIN CRAIN'S OFFICE—DAY

Mr. Crain is seated behind his desk, doing paperwork. He turns to the CAMERA.

MR. CRAIN

What you've just seen was a dramatization based on fact. It was designed to emphasize the importance of a strong governmental affairs department in today's extremely competitive business environment. As we're all well aware, the telecommunications industry is hardly the secure and controlled environment it used to be. Today, we need every advantage we can gain on both a business and a political level, and our present govern-

DLTV - 225A 10

mental affairs group provides us with that critical advantage. I support their activities 100%, and I hope I can count on each of you for that same support. Thank you. . . .

FADE-OUT

THE END

APPENDIX

4

Budget

❑ ❑ ❑ ❑ ❑

PRODUCTION BUDGET

TITLE: GOVERNMENTAL AFFAIRS: A "TYPICAL" DAY PROJECT#: 225A

PRODUCER: A. SPARKS DATE: 8/15/92

100	CONCEPT/SCRIPT:	QTY	HRS/ DAYS	UNIT COST	LOADED RATE	SUB-TOTAL	COMMENTS
101	PRODUCER	1	5	$360	$360	$1800	
102	SCRIPT/PROPOSAL	1	1	$250	$302	$302	
103	SCRIPT/TREATMENT				$0	$0	
104	SCRIPT/WRITING	1	1	$1800	$2173	$2173	
105	SCRIPT/REWRITE				$0	$0	
PREPRODUCTION:							
252	SET CONST.				$0	$0	
253	SET RENT	1	1	$500	$500	$500	
254	OTH/PP/E				$0	$0	
280	STAGE B&L	1	10	$100	$100	$1000	HOURLY RATE
281	OTH/PP/F			$100	$100	$0	HOURLY RATE
PRODUCTION:							
302	DIRECTOR	1	10	$500	$604	$6035	
303	AD/AP	1	9	$250	$302	$2716	
304	TD	1	2	$300	$362	$724	
305	VIDEO ENG				$0	$0	
306	VIDGPHER	1	6	$375	$453	$2716	
307	CAMERA (STUDIO)	2	2	$200	$241	$966	
308	AUDIO ENG	1	5	$250	$302	$1509	
309	TELEPROM				$0	$0	
310	GRIP (B&L)	1	2	$200	$241	$483	
311	GAFFER	1	6	$225	$272	$1629	
312	COMPU GRPHC OPER				$0	$0	
313	PA	1	9	$100	$121	$1086	
314	STIL/PHOTO				$0	$0	
315	TALENT/COMP			$45	$45	$0	HOURLY RATE
316	TALENT/SPOKESPER				$0	$0	
317	TALENT/DAY PLYR	7	1	$348	$460	$3218	
318	TALENT/VO				$0	$0	
319	TALENT/EXTRA	5	2	$100	$132	$1321	
320	OTH/PROD/TIME				$0	$0	
321	CG OPERATOR				$0	$0	

PRODUCTION (CONT.):	QTY	HRS/ DAYS	UNIT COST	LOADED RATE	SUB-TOTAL	COMMENTS
327 PERMITS	1	1	$200	$241	$241	
333 PRESENTATION			$45	$45	$0	HOURLY RATE
350 MESSENGER				$0	$0	
351 TRAVEL/MEALS	10	5	$9.5	$10	$475	
353 CAMERA RENT	1	3	$750	$750	$2250	
354 LIGHT RENT				$0	$0	
355 AUDIO RENT	1	3	$176	$176	$528	
356 VTR RENT				$0	$0	
357 VEH RENT				$0	$0	
375 VAN			$50	$50	$0	DAILY RATE
382 STAGE COMPLETE	1	10	$125	$125	$1250	HOURLY RATE
383 AUDIO RECORD FAC	1	10	$40	$40	$400	HOURLY RATE
384 COMPUTER GRPHC			$100	$100	$0	HOURLY RATE
385 REM EQUIP			$40	$40	$0	HOURLY RATE
386 OTH/PROD/FAC			$100	$100	$0	
387 CAM-HI			$30	$30	$0	HOURLY RATE
388 CAM-NW			$40	$40	$0	HOURLY RATE
389 CHR. GEN			$40	$40	$0	HOURLY RATE
390 STK FTG-SEARCH/DUB			$55	$55	$0	HOURLY RATE
391 STK FTG-PER MIN			$50	$50	$0	MIN RATE
POSTPRODUCTION:						
403 EDITOR—OFLN	1	5	$275	$332	$1660	
404 EDITOR—ONLN	1	2	$350	$422	$845	
405 PROD DUPE OPR	1	6	$45	$45	$270	HOURLY RATE
406 SWEETENING ENG	1	1	$275	$332	$332	
409 DIST. EQUIP. REPAIR			$45	$45	$0	HOURLY RATE
410 OUTSIDE POST/DUPING				$0	$0	
411 OTHER POSTPROD EXP				$0	$0	
412 EVALUATIONS			$45	$45	$0	HOURLY RATE
413 DIST. DUPER			$45	$45	$0	HOURLY RATE
414 MAT. & SHIP	1	1	$200	$200	$200	
454 MUSIC FEES (BLANKET)				$0	$0	
455 RENT OFLN				$0	$0	
456 RENT ONLN				$0	$0	
482 AUDIO SWEET	1	8	$50	$50	$400	HOURLY RATE
484 ONLN-1-A	1	16	$110	$110	$1760	HOURLY RATE
485 ONLN-1-B			$110	$110	$0	HOURLY RATE
486 ONLN-2-A			$110	$110	$0	HOURLY RATE
487 ONLN-2-B			$110	$110	$0	HOURLY RATE
488 OFLN-A BAY	1	40	$35	$35	$1400	HOURLY RATE
489 OFLN-B BAY			$35	$35	$0	HOURLY RATE
490 OFLN-C BAY			$35	$35	$0	HOURLY RATE
493 DUPE DIST. FACILITY			$30	$30	$0	HOURLY RATE
496 DUPE PRD. FACILITY	1	6	$30	$30	$180	HOURLY RATE
501 MARKETING LABOR			$30	$30	$0	HOURLY RATE
502 MARKETING EXPENSES				$0	$0	
				SUBTOTAL	$40,367.76	
				SUBTOTAL	$40,367.76	
360 OVERHEAD				5%-OH	$2,018.39	
				TOTAL	$42,386.15	

APPENDIX

5

Scene Breakdown—Chronological

❑ ❑ ❑ ❑ ❑

①

SCN	MASTER DESCRIPTION	INT./ EXT.	PRINCIPLES	EXTRAS	SET-UPS	PAGE(S)	PROPS	SPECIAL EQUIP.	OTHER CONSIDERATIONS	SHOOT HRS.
1	D&C H.Q. Build LS. Estab	E	–	–	1	1	–	–	People enter/exit?	.5
2	Facilitator ends class. Students exit	I	3	5	3	1	–	Dolly	Wardrobe- mgmt & hourly	2.5
3	Tom & Phylis get coffee, MT & talk -brk room	I	2	5	3 cam. live	1-4	cups	–	Set - all elem. 1st set up	2.5
4	Flashback - Director drops work on Phylis	I	2	Empl.	3	4	"Bill"	–	Do hand-held?	2.0
5-1	Montage -1 Phylis on phone	I	1	Empl.	3	4	–	–	Shoot several angles for different times/situations	.5
5-2	Phylis meets person in legal dept.	I	1	1 Empl.	1	4	notebook w/ paper	–	Motion/hand-held	1.5
5-3	Const. Dept. Phylis meets w/ Supt.	I or E	1	1 Empl. Supt.	2	4	as above	–	Important same Supt. appears later & has lines	2.0
5-4	Phylis on phone w/ Sacramento rep.	I	2	Empl.	2	5	pos. paper	–	Shoot Sac. rep. @ same loc.	2.0

(NOTE ITEMS TOO LENGTHY TO LIST ON ATTACHMENT OR REVERSE SIDE)

13.5

Figures A5.1–A5.3 Chronological scene breakdown lists production elements in script order (note scene numbering in left column). It requires a good deal of analysis by the director and provides the first step in arriving at a shooting schedule.

(2)

SCN	MASTER DESCRIPTION	INT./ EXT.	PRINCIPLES	EXTRAS	SET-UPS	PAGE(S)	PROPS	SPECIAL EQUIP.	OTHER CONSIDERATIONS	SHOOT HRS.
6	Brk. Room Continuation of Scn. 3	I	2	5	3 cam. live	5-6	cups	–	Set	.5
7	Phylis desk Rep. passes, says mtg.	I	2	Empl.	1	6	paper pencil notebook	–	hand-held ?	1.5
8	Mtg. in conf. room Director & Phylis & Reps.	I	3	4	3 cam- live	6-7	–	–	1st scn. Studio set up	2.5
9-1	Montage 2 Phylis on phone & typing & calling	I	1	Empl.	1	7	pizza coke coffee	–	hand-held or dolly Poss. split scrn.	.5
9-2	Const. Dept. Superint. talks to Phylis	I or E	1	Empl	1	7	–		Possible split scrn w/9-1	1.5
9-3	Phylis back @ legal	I	1	1 Empl.	1	7	NT. BK.	–	Motion	.5
10	Brk. Room continues from Scn. 6	I	2	5	3 cam	8	cups	–		.5
11	Flashback Director's office Reads letter	I	3	4	3 cam. (scn. 8)	8-9	letter	–	New wardrobe same set as scn - 8	.5

(NOTE ITEMS TOO LENGTHY TO LIST ON ATTACHMENT OR REVERSE SIDE)

21.5

(3)

SCN	MASTER DESCRIPTION	INT./ EXT.	PRINCIPLES	EXTRAS	SET-UPS	PAGE(S)	PROPS	SPECIAL EQUIP.	OTHER CONSIDERATIONS	SHOOT HRS.
12	Brk. Room Cont. Scn. 10 Exits	I	2	5	3 cam.	9-10	cups	–		.5
13	Tom calls Supt. turns out to be same	I	2	–	2	11	–	–	split screen ?	1.5
14	C.E.O. Do in studio	I	1	–	1	10-11	–	–		.5
									Notes: Watch wardrobe	24
									Movement in montage sens.	
								initial	Travel - Lunches, etc. studio set up	10.5
									TOTAL →	34.5
									3½ days	

(NOTE ITEMS TOO LENGTHY TO LIST ON ATTACHMENT OR REVERSE SIDE)

APPENDIX

6 Scene Breakdown—Reorganized

SCN	MASTER DESCRIPTION	INT./ EXT.	PRINCIPLES	EXTRAS	SET-UPS	PAGE(S)	PROPS	SPECIAL EQUIP.	OTHER CONSIDERATIONS	SHOOT HRS.
3	Tom & Phylis get coffee, MT & talk - brk. room	I	2	5	3 cam. live	1-4	cups	–	Set - all elem. 1st set up	2.5
6	Brk. Room Continuation of Scn. 3	I	2	5	3 cam live	5-6	cups	–	Set	.5
10	Brk. Room continues from scn. 6	I	2	5	3 cam.	8	cups	–		.5
12	Brk. Room cont. Scn. 10 Exits	I	2	5	3 cam	9-10	cups	–		.5
8	Mtg. in conf. room. Director & Phylis & Reps	I	3	4	3 cam - live	6-7	–	–	1st scn. studio set up	2.5
11	Flashback Director's office Reads letter	I	3	4	3 cam. (scn. 8)	8-9	letter	–	New wardrobe same set as scn - 8	.5
14	C.E.O. Do in Studio	I	1	–	1	10-11	–	–		.5

(NOTE ITEMS TOO LENGTHY TO LIST ON ATTACHMENT OR REVERSE SIDE)

Day 1 - 7.5 studio

Figures A6.1–A6.4 Chronological scene breakdown is cut into strips and arranged not according to the script but in a cost-effective *shooting* order. Note that the director had also marked scene numbers to highlight similarities. For instance, scenes 3, 6, 10, and 12 are all to be shot on the studio break room set with Tom and Phylis; thus, the director marked each with a square. This marking was done prior to cutting the breakdown into strips. Shooting order breakdown is the second step in developing a shooting schedule.

SCN	MASTER DESCRIPTION	INT./ EXT.	PRINCIPLES	EXTRAS	SET-UPS	PAGE(S)	PROPS	SPECIAL EQUIP.	OTHER CONSIDERATIONS	SHOOT HRS.
4	Flashback - Director drops work on Phylis	I	2	Empl.	3	4	"Bill"	–	Do hand-held ?	2.0
5-1	Montage -1 Phylis on phone	I	1	Empl.	3	4	–	–	Shoot several angles for different times / situations	.5
5-4	Phylis on phone w/ Sacramento rep.	I	2	Empl.	2	5	pos. paper	–	Shoot Sac. rep. @ same loc.	2.0
7	Phylis desk Rep. passes, says mtg.	I	2	Empl.	1	6	paper pencil notebook	–	hand-held ?	1.5
9-1	Montage 2 Phylis on phone & typing & calling	I	1	Empl.	1	7	pizza coke coffee	–	hand-held or dolly Poss. split scrn.	.5
1	D & L H.Q. Build LS. Estab.	E	–	–	1	1	–	–	People enter/exit ?	.5

(NOTE ITEMS TOO LENGTHY TO LIST ON ATTACHMENT OR REVERSE SIDE)

DAY 2

-7.0
-2 LOCATIONS

5-2	Phylis meets person in legal dept.	I	1	1 Empl.	1	4	notebook w/paper	–	Motion/hand-held?	1.5
9-3	Phylis back @ legal	I	1	1 Empl.	1	7	NT. BK.	–	Motion	.5
9-2	Const. Dept superint. talks to Phylis	I or E	1	Empl.	1	7	–		Possible split scrn w/9-1	1.5
5-3	Const. Dept. Phylis meets w/Supt.	I or E	1	1 Empl. Supt.	2	4	as above	–	Important same supt. appears later & has lines	2.0
13	Tom calls Supt. turns out to be same	I	2	–	2	11	–	–	Split screen?	1.5

DAY 3
- 3 MOVES
- EXT. LOCATIONS
7.0

2	Facilitator ends class. Students exit.	I	3	5	3	1	–	Dolly	Wardrobe - mgmt & hourly	2.5

Day 4

½ Day

APPENDIX

7 Shooting Schedule

SHOOTING SCHEDULE
FORM PR 2523

V No. 225A TITLE GOVERNMENTAL AFFAIRS

DISTRIBUTION:
White - Producer
Yellow - Director
Pink - Ad
Goldenrod - Project File

PAGE: 1

PRODUCER ANGELA DIRECTOR JOHN AD JENNY

DATE	DAY	TIME	SCENE	PAGE(S)	DESCRIPTION	LOCATION	TALENT(S)	PROPS
9/21	M	7:00	-	-	Crew call time. Talent call time. Set up for scene 3	STUDIO	Oscar Ellen	coffee cups
"	"	8:00	3	1-4	Tom and Phylis enter break room	"	"	"
"	"	9:45	6	5-6	2nd break room scene Tom and Phylis cont.	"	"	"
"	"	10:15	10	8	3rd break room scene	"	"	"
"	"	11:00	12	9-10	4th and final break room scene	"	"	"
"	"	11:30	-	-	Lunch			
"	"	12:30	-	-	Talent call - Paul and extras Set up for scenes 8&11	"	Paul Mollet & extras	paperwork & memo
"	"	1:30	8	6-7	Conference room meeting Director, Phylis and staff	"	"	"
"	"	3:00	11	8-9	2nd conference room scene - CEO memo	"	"	"
"	"	4:00	-	-	Set up for CEO close	"	CEO	-
"	"	4:30	-	-	CEO			
"	"	5:00	-	-	Wrap			

Figures A7.1–A7.4 Finalized shooting schedule is the culmination of the director's preproduction work. A great deal of knowledge and analysis have gone into its development.

SHOOTING SCHEDULE
FORM PR 2523

DISTRIBUTION:
White - Producer
Yellow - Director
Pink - Ad
Goldenrod - Project File

V No. 225A TITLE GOVERNMENTAL AFFAIRS

PRODUCER ANGELA DIRECTOR JOHN AD JENNY PAGE: 2

DATE	DAY	TIME	SCENE	PAGE(S)	DESCRIPTION	LOCATION	TALENT(S)	PROPS
22	T	7:00	-	-	Crew and cast call time Load equipment & move to Gov. Affairs	-	Paul, Ellen Reps, G. Mintz Jane	papers in sheath
"	"	7:30	-	-	General set-up	Gov. Affairs	"	"
"	"	9:00	4	4	Director passes Phylis and gives her paperwork	"	Paul Ellen	"
"	"	10:30	5-4	5	Phylis talks to Sacramento Rep.	"	Ellen Gerald	
"	"	11:30	-	-	Pre-light/set for 9-1, 5-1, &7	"	-	-
"	"	12:30	-	-	Lunch	"		
"	"	1:30	5-1	4	Phylis on phone	"	Ellen	
"	"	2:30	7	6	Rep. passes Phylis' desk	"	Ellen Jane	
"	"	3:30	9-1	7	Phylis on phone, typing, writing		Ellen	
"	"	4:00	-	-	Strike location	-	-	-
"	"	5:00	1	1	EXT. D&L - John and Bill only crew strikes	-	-	-
"	"	5:30	-	-	Wrap			

SHOOTING SCHEDULE
FORM PR 2523

DISTRIBUTION:
White - Producer
Yellow - Director
Pink - Ad
Goldenrod - Project File

V No. 225A TITLE GOVERNMENTAL AFFAIRS PAGE: 3

PRODUCER ANGELA DIRECTOR JOHN AD JENNY

DATE	DAY	TIME	SCENE	PAGE(S)	DESCRIPTION	LOCATION	TALENT(S)	PROPS
23	W	7:00	–	–	Crew and cast call- Load trucks	–	Ellen	
"	"	8:00	–	–	Set up for 5-2, 9-3	Legal Dept.	Ellen & Empl.	
"	"	9:30	5-2	4	Phylis visit Legal, meets empl.	"	"	papers notebook
"	"	10:00	9-3	7	Phylis returns to legal	"	Ellen Jane Emp.	"
"	"	10:30	–	–	Strike and leave for Const.			
"	"	11:30	–	–	Lunch			
"	"	12:30	–	–	Set up for sc. 9-2 Call time for Paul M.	Const.		
"	"	1:30	9-2	7	Supt. on phone to Phylis	"	Paul Musser	
"	"	2:00	–	–	Set up for sc. 13	"		
"	"	3:00	13	11	Supt. talks to Tom	"	Paul M.	
"	"	3:30	–	–	Set up for 5-3			
"	"	4:30	–	5-3	Phylis meets Supt.		Paul Ellen	

SHOOTING SCHEDULE
FORM PR 2523

V No. 225A TITLE GOVERNMENTAL AFFAIRS

PRODUCER ANGELA DIRECTOR JOHN AD JENNY

DISTRIBUTION:
White - Producer
Yellow - Director
Pink - Ad
Goldenrod - Project File

PAGE: 4

DATE	DAY	TIME	SCENE	PAGE(S)	DESCRIPTION	LOCATION	TALENT(S)	PROPS
24	Th	8:00	–	–	Crew and Cast call		Ellen, Oscar Derk, extras	
"	"	8:30	–	–	Set up for E.I. meeting	Conference Rm	"	
"	"	10:00	2	1	E.I. class adjourns	"	"	
"	"	12:00	–	–	Lunch			
"	"	1:00	–	–	Wrap			

APPENDIX

8 Shot List

❑ ❑ ❑ ❑ ❑

Shot List—Governmental Affairs

Scene Number	*Shot*
1	HQ Build— WS Zoom in Zoom out
2	Facilitator ends class, students exit WS, from rear of room Dolly, from rear to reveal exit Singles—students 2 & 3 shots—students
3	Live switched
4	Phylis and director at desk 2 shot—pan and push to director Single Phylis—and reaction CU paperwork hits desk
5.1	Phylis on phone MCU—dead on WS—high— + CU & zoom OTS paperwork
5.2	Phylis at the legal dept. WS Phylis enters, shakes hands w/legal employee MS—same action ECU—hands meet and shake CU Phylis OTS CU paperwork being shared
5.3	Phylis w/ superintendent in field 2 shot Single supt. Single Phylis
5.4	Phylis on phone to Sacramento 3/4 MCU CU notes OTS notes (Split screen?)
6	Switched live

Shot List—*(continued)*

Scene Number	*Shot*
7	Phylis is told of meeting MS—pull or dolly out as rep approaches CU rep CU reaction—Phylis (Hand-held?)
8	Switched live
9.1	Phylis typing, calling WS—MS—low MCU 3/4 (for split screen) ECU fingers OTS PC scrn/keyboard
9.2	Phylis phone conv. w/ superintendent CU dolly back MS lock off (Split screen)
9.3	Phylis 2nd visit to legal 2 shot—Phylis/rep Single Phylis Single rep
10	Switched live
11	Switched live
12	Switched live
13	Tom calls superintendent MS Tom at pay phone CU same ECU Tom dialing (shoot to reveal split screen) MS supt on phone MCU same (Shoot to reveal split screen)
14	CEO closing message MCU

Glossary

"action" The director's cue to crew and actors that the scene should begin.

alternate An alternate version, usually of a master scene, to provide the director with a different perspective on the action, dialogue, or both.

ambience The background noise present at any location. Room ambience is also called room *tone.* Ambience on a street corner would include traffic sounds such as horns honking and perhaps pedestrian chatter.

angle on A camera term noting that the angle should be on a person or thing but not specifying a focal length (such as medium close-up or close-up).

apple box A multipurpose hollow wooden cube used in production as stops, wedges, platforms, etc. Various sizes are called *full apple, half apple,* and *quarter apple.*

audio The sound portion of a script—narration, dialogue, sound effects, and music. In the column format, the sound or audio column is usually written on the right side of the page. In the screenplay format, it appears both in the scene descriptions and in the dialogue or narration column down the center of the page. The term *audio* also refers to the sound portion of a tape or film recording.

audiovisual (AV) Any type of sound (audio) and picture (visual) presentation. AV is usually used to describe a slide or overhead presentation, rather than a film or videotape production.

BG A script abbreviation for *background.*

big shot A very wide, establishing shot.

blocking Actors' movements on the set in relation to the camera. Blocking should appear motivated to look natural.

"buy" A director's term of acceptance for a shot. The *buy take* is the best of the takes recorded and most likely the one the editor will use in the show. It is noted in the master script with a circle around the take number. Also called the *circled take.*

character acting An acting discipline in which the actor bases the performance on the personality makeup of a character. The immediate and distant personal history of the character often comes into play. For instance, an actor playing a depressed executive might want to know about the character's childhood, education, job history, and parents, to bring credibility to the role.

client The person requesting, approving, and often paying for the program. As such, the client is one of the director's key business contacts. The other key contact is the producer.

close-up (CU) A shot of an actor or item that fills the majority of the screen. In a close-up of a female character, for instance, we would see her entire face and the upper part of her shoulders. In a close-up of a glass, we would see the entire glass and a small area of the surrounding surface it was on.

column format A script format in which the sound and picture elements are separated into two columns on the page. Usually, the picture column is on the left and the sound column is on the right. Often used by television directors because of its ease of use in multicamera production.

concept The basic premise on which the script, or segments of it, will be based. Concepts link content and design with visualization and story line.

content expert A person who is an authority or expert on the subject about which the director is shooting. Also called a *subject matter expert* (SME) or a *technical adviser.*

corporate culture The basic philosophy or standards system that a large company or corporation uses as a framework for its day-to-day business activities. For instance, a corporate culture may be conservative, in which case all management employees might be required to wear certain kinds of business attire.

corporate politics Personal preferences and maneuvers that affect what should be strictly business decisions. For instance, a finance manager might not like a peer in the accounting department; therefore, he rejects a valid proposal (which could be production related) presented by her.

corporate television Television productions produced for an in-house audience of company employees. Also called *industrials* or *business productions.*

coverage A term referring to the amount and type of footage shot to allow a scene to be edited properly. Standard coverage typically consists of a wide shot of all action, medium shots of most portions of the same, action, close-ups of parts of the action that need emphasis, and extreme close-ups or inserts of action, such as a number being dialed on a telephone or a specific key being struck on a computer.

crab A sideways dolly movement, usually perpendicular to the original direction of travel.

cut An instantaneous picture change used as a transition between scenes. Also used within scenes to facilitate viewing the action from different angles.

dialogue Conversation spoken between at least two actors in a role-play situation.

director The person responsible to the producer to analyze and execute the written script into film, audio, digital, or video footage appropriate for editing into a completed program.

dissolve A transitional effect in which one picture fades out as another fades in. Most often used to suggest a passage of time or a major location change. Also used to "soften" jarring or abrupt cuts.

dolly A forward or backward movement of the camera during a shot. This is opposed to a *truck,* which is a left or right sideways camera motion. The terms *truck* and *dolly* are, however, often used synonymously.

drafts Versions of the script. Typically, a scriptwriting project goes through about three drafts. The first draft is the writer's initial version. This is given to the producer and sent to the client. Revisions are then worked into the second draft. The third draft is often considered the final draft or shooting script.

DVE Acronym for *digital video effects.* Often used in scripts to precede a description of a digital effect. For example: DVE—PICTURE FLIES OFF SCREEN LEFT.

entrance An actor's cue to move on stage or into the camera's frame.

establishing shot Usually the first wide shot shot of an environment. Sometimes written this way: WIDE SHOT—TO ESTABLISH.

exit An actor's cue to leave the stage or camera's frame.

EXT Abbreviation for *exterior.* Used in scene headings.

extreme close-up (ECU) An extremely close camera focal length. In an extreme close-up of a face, we would see only a part of the entire face. The lips or eyes, for instance, might fill the entire screen.

fade in/fade out Editing terms used to begin and end a script or note a major transition. The word *fade* means that the picture fades in from black or some other color or fades out to that color.

FG A script abbreviation for *foreground.*

flag A black-cloth, wire-rimmed, fanlike object used to shade light from an object or actor.

flat fee A payment arrangement in which the director is offered a preset amount of money for a project, regardless of the time required to prepare and shoot it.

focus puller A camera assistant who changes focus on the fly as the camera moves during recording.

freeze frame A picture that is stopped at some key spot, often for dramatic effect, such as the freeze frame of an athlete crossing the finish line in a race. Also used to create a still image background panel for superimposed titles.

high angle A camera angle shot from above, looking down on the action.

HMI An acronym referring to a *halogen metal iodide* light, a production light that creates bluish light and is used to mix with or "create" sunlight or moonlight.

host An on-camera spokesperson. The host talks directly to the audience (camera).

Insert Usually a close-up or extreme close-up of an object that is inserted into the main action. For example, a character may begin dialing the telephone in a wide shot. An insert of her fingers turning the dial is cut in so that we are able to recognize the number she is calling. When the dialing is completed and the character begins to talk, we return to the wide shot or some other angle.

Instructional design A method of custom-designing instructional material based on such criteria as audience factors, objectives, the problem to be solved, and program utilization. Design documents, such as a program needs analysis, are sometimes given to the director with the script at the outset of a project.

INT Abbreviation for *interior.* Used in scene headings.

key Short for *chromakey.* Used to denote the combining of pictures or a picture and title. For example: KEY IN LOGO OVER SHOT or KEY TITLES. Sometimes used synonymously with *super,* meaning *superimpose.*

limbo A seamless, usually featureless background, often of one color. Used as a generic set environment. A spokesperson or a vignette using role-play actors might be set in limbo. This is usually done for a dramatic effect or because it is cheaper than building a set.

"the line" An imaginary line that runs through any person or object in motion on the screen. This line establishes screen direction, meaning the logical direction of motion or attention a viewer becomes oriented to. If the line is crossed when shooting, the screen direction of the person or object changes, often disorienting the viewer. Also called *the stage line* or *the screen direction line.*

low angle A camera angle shot from below, looking up at the action.

master script The script brought to the field on which the production notes are taken during the shoot. Later, it is used by the editor to locate the circled takes or "buys" of each shot. It is often filed and retained when the program is completed.

master shot A complete segment of action shot uncut in one pass. Often, a master shot will be a wide focal length containing all actors and showing all action.

medium close-up (MCU) A camera focal length that frames an individual's face, starting at the chest.

medium shot (MS) A camera focal length in which the majority of a person's body can be seen. A medium shot is framed roughly from the thighs up. A *medium two shot* would include two people in about the same focal length.

method acting A school of acting that relies on detailed sensory memories recalled from an actual past experience and applied to a scene being played in the present. For instance, an actor might recall the smells, temperature, touches, voices, sounds, and textures present in her dentist's office to accurately portray a character who is anxious about having a tooth pulled.

montage A series of images, often fast paced and cut to music, used to suggest a compression of time. Also used for dramatic effect, many times in the openings of programs.

multicamera A form of production in which multiple cameras are switched on the fly to edit the scene as it plays out. News shows and sports events are examples of live multicamera productions. Soap operas are also shot multicamera style, but they are recorded on videotape.

music up/music under/music out Script notations. *Music up* denotes a music increase in volume; *music under,* a decrease to allow other sounds more prominence; and *music out,* a complete fade-out.

narration Words spoken by a narrator or host. Narration is spoken directly to the audience (camera).

narrator An off-camera spokesperson. A host is an on-camera version of a narrator.

net A wire-rimmed, fanlike object with material similar to nylon stockings stretched taught across it. Used to shade an actor or object.

OC or OS Abbreviations for *off-camera* or *off-scene,* meaning generally the same thing as voice-over (VO)—a voice is heard while the speaker is not being seen. Usually used in the dialogue or narration heading.

off-line edit The first editing session, in which a director's work is assembled into a complete program. Also called the *rough cut.* Low-quality work prints of the original footage are used to allow decision making involving timing, pace, and the inclusion or deletion of certain shots. The director is usually involved in the off-line edit, and the client and producer typically approve and have input on it before the final, on-line edit is performed.

on-line edit The final edit, in which music, effects, titles, and the final "tightening" and polish are applied. The on-line edit is based primarily on prior decisions made in the off-line edit. Also called the *fine cut.*

over the shoulder (OTS) A camera angle looking over one person's shoulder at another person. "OTS—Jill" would mean we are looking at Jill over someone else's shoulder.

pan A horizontal rotation of the camera on its head. Short for *panorama.*

ped up or ped down A rising or lowering of the camera on its head. Short for *pedestal.*

POV A script abbreviation for *point of view.*

"prep" Jargon for *preparation* or *preproduction*—the period of time given to the director to prepare for the shoot. Also refers to the work that goes on during preproduction.

producer The person responsible and accountable for all aspects of a film, video, digital, or audio production—from script development through shooting and editing.

rack focus A drastic change of focus to create emphasis or change the viewer's perspective. For example, the camera might rack focus from a street sign in the foreground to a man walking in the distance.

reveal A zoom, pan, or other camera move that reveals something in the frame the viewer had not previously been aware of. For example: ZOOM OUT TO REVEAL John entering the office.

reverse angle A view that is 180 degrees from the preceding shot. For example, if we are looking over the shoulder of a driver moving down the street in a car, a reverse angle would be the view looking directly backward out the rear window.

rewrite A major overhaul of an existing script.

role-play A scene designed to dramatize a situation. Can be serious or humorous.

"roll tape" A director's cue to the technical director or camera operator. When a director says, "Roll tape," it signals the crew that he is satisfied that the scene is now ready to be recorded.

rough cut See **off-line edit.**

screenplay format A script format in which the scene descriptions are written across the entire page and the sound is written in a column down the center. The standard "Hollywood" format. Most often used for single-camera shooting.

screen right/screen left The position of something in the scene as viewed from the camera's perspective. The direction is the opposite from anyone on the set who is looking back at the camera. An actor's right, when looking at the camera, is screen left.

script The framework for an audiovisual production. The script is a precisely written, visual description of all program elements that is then used by the director and producer to create the program.

shooting script The final draft, with all changes made. A production crew takes the shooting script to the field.

shock cut A cut used to dramatize a transition and shock the audience. For example, as a gun is about to be fired at a male protagonist, a script might call for a shock cut to an extreme close-up of a telephone ringing in his wife's office.

silk Sheetlike material, stretched over a square or rectangular frame and mounted over a scene or actor. Used to diffuse harsh shadows, often from sunlight.

single One person only seen in a medium close-up or medium shot.

single camera A form of production in which a single camera is moved to different positions and the scene, or parts of it, is recorded several times from various perspectives. This "coverage" is then edited together later.

slate A small board placed in front of the camera at the "head" of each scene. The slate gives that scene its own informational "signature." This allows the editor to later choose the proper scenes for inclusion in his editing. The slate usually contains the name of the project, a project number, the date, the video or film reel, the scene number, the take number, and the director's and camera operator's names.

SOT Abbreviation for *sound on tape.* Used to indicate the fact that sound is coming from some prerecorded tape source, rather than live—as it might be spoken by an actor or narrator.

sound effect Notation written into a script to call for a sound effect. Also called or written in as *SFX*.

"speed" A verbal signal to the director from the technical director or camera operator to let him know that recording has begun and he may now give the "Action" cue.

sting A short piece of music used to punctuate a transition or heighten a dramatic moment.

stock footage Footage pulled from a library or archive to be included in a program.

story outline Another term for *treatment*. A scene-by-scene description of the story, written in simple narrative terms.

super Short for *superimposition*. Refers to something superimposed over a picture. For example: SUPER OPENING TITLES.

talent Jargon for *actors*.

technical adviser A subject matter expert who is usually present on the set.

technical director (TD) The technician responsible for the electronic aspects of a studio recording. The TD "switches" or operates the production switcher, which controls camera signals that are recorded or broadcast. He also controls video levels, sets up, and monitors all studio equipment associated with the shoot.

tight shot Using a tight camera frame. For instance: TIGHT TWO SHOT.

tilt up/tilt down A vertical rotation of the camera on its head.

treatment A scene-by-scene narrative describing the story and visual aspects of a script that often is yet to be written. Used as a pre-script-approval step to be sure clients and producers like the concept the writer has come up with.

truck A movement of the camera in a sideways motion—right or left. This is opposed to a *dolly*, which is a movement forward or backward. *dolly* is often used interchangeably with *truck*, however.

tungsten A production light that creates an orange-colored light used to mix with or "create" interior room light.

two shot A shot of two actors, positioned close together, usually in conversation.

upstage/downstage Terms referring to areas in relation to the camera. *Upstage* is the area in the background, farthest from the camera. *Downstage* is in the foreground, closest to the camera.

video The picture column in a two-column script format. Also refers to the video portions of recorded programs and is used loosely to refer to complete programs and the entire field.

vignette See **role-play**.

voice-over (VO) Portions of narration or dialogue heard while the viewer is seeing something other than the person speaking. Means a voice heard "over" pictures.

wide shot (WS) A camera focal length encompassing the entire scene or a large part of it. Often used to establish a room or environment in which the scene is taking place.

wipe A transitional effect in which one picture is wiped off the screen by another. Can take place horizontally, vertically, or in circular or rectangular patterns.

zoom in/zoom out A change of focal length by the camera. For example, from a wide shot, the script might call for a "ZOOM IN TO CU—JENNY."

Bibliography

Directing

Bare, Richard L., *The Film Director.* New York: Collier Books, Macmillan Publishing Company, 1971.

Dmytryk, Edward, *On Screen Directing.* Stoneham, MA: Focal Press, 1983.

Kennedy, Thomas, *Directing Video.* White Plains, NY: Knowledge Industry Publications, Inc., 1989.

Rabiger, Michael, *Directing the Documentary.* Stoneham, MA: Focal Press, 1987.

Rabiger, Michael, *Directing: Film Techniques and Aesthetics.* Stoneham, MA: Focal Press, 1989.

Sherman, Eric, *Directing the Film: Film Directors on Their Art.* Los Angeles: Acrobat Books, 1976.

Scriptwriting

DiZazzo, Ray, *Corporate Scriptwriting: A Professional's Guide.* Stoneham, MA: Focal Press, 1991.

Swain, Dwight V. with Joyce R. Swain, *Film Scriptwriting: A Practical Manual,* 2nd Ed. Stoneham, MA: Focal Press, 1988.

Van Nostran, William J., *The Scriptwriter's Handbook.* White Plains, NY: Knowledge Industry Publications, Inc., 1989.

Acting

Blum, Richard A., *Working Actors: The Craft of Television, Film, and Stage Performance.* Stoneham, MA: Focal Press, 1989.

Dmytryk, Edward and Jean Porter, *On Screen Acting.* Stoneham, MA: Focal Press, 1984.

Index